PERGAMON INTERNATIONAL LIBRARY
of Science, Technology, Engineering and Social Studies
*The 1000-volume original paperback library in aid of education,
industrial training and the enjoyment of leisure*
Publisher: Robert Maxwell, M.C.

The Analysis of Response in Crop and Livestock Production

THE PERGAMON TEXTBOOK
INSPECTION COPY SERVICE

An inspection copy of any book published in the Pergamon International Library will gladly be sent to academic staff without obligation for their consideration for course adoption or recommendation. Copies may be retained for a period of 60 days from receipt and returned if not suitable. When a particular title is adopted or recommended for adoption for class use and the recommendation results in a sale of 12 or more copies, the inspection copy may be retained with our compliments. If after examination the lecturer decides that the book is not suitable for adoption but would like to retain it for his personal library, then a discount of 10% is allowed on the invoiced price. The Publishers will be pleased to receive suggestions for revised editions and new titles to be published in this important International Library.

Some other titles of interest

BUCKETT, M. Introduction to Animal Husbandry

CLARK, C. The Value of Agricultural Land

CLAYTON, E. S. Agrarian Development in Peasant Economies

COWLING, K. et al. Resource Structure of Agriculture

DODSWORTH, T. L. Beef Production

DOUGLASS, R. W. Forest Recreation

GARRETT, S. D. Soil Fungi and Soil Fertility

GILCHRIST SHIRLAW, D. W. A Practical Course in Agricultural Chemistry

KENT, N. L. Technology of Cereals (with special reference to wheat)

LAWRIE, R. A. Meat Science

LOCKHART, J. A. R. & WISEMAN, A. J. L. Introduction to Crop Husbandry

MILLER, R. & A. Successful Farm Management: The Bannockburn Conservation Group System for Practical Book-keeping and Planning

NELSON, R. H. An Introduction to Feeding Farm Livestock

PARKER, W. H. Health and Disease in Farm Animals

PRESTON, T. R. & WILLIS, M. B. Intensive Beef Production

ROSE, C. W. Agricultural Physics

SHIPPEN, J. M. & TURNER, J. C. Basic Farm Machinery

TAYLOR, J. A. The Role of Water in Agriculture

VOYSEY, A. Farm Studies

YEATES, N. T. M., EDEY, T. N. & HILL, M. K. Animal Science

The terms of our inspection copy service apply to all the above books. Full details of all books listed will gladly be sent upon request.

The Analysis of Response in Crop and Livestock Production

SECOND EDITION

by

JOHN L. DILLON

Professor of Farm Management
University of New England, Armidale, Australia

PERGAMON PRESS

OXFORD · NEW YORK · TORONTO · SYDNEY
PARIS · FRANKFURT

U. K.	Pergamon Press Ltd., Headington Hill Hall, Oxford OX3 0BW, England
U. S. A.	Pergamon Press Inc., Maxwell House, Fairview Park, Elmsford, New York 10523, U.S.A.
CANADA	Pergamon of Canada Ltd., 75 The East Mall, Toronto, Ontario, Canada
AUSTRALIA	Pergamon Press (Aust.) Pty. Ltd., 19a Boundary Street, Rushcutter Bay, N.S.W. 2011, Australia
FRANCE	Pergamon Press SARL, 24 rue des Ecoles, 75240 Paris, Cedex 05, France
WEST GERMANY	Pergamon Press GmbH, 6242 Kronberg-Taunus, Pferdstrasse 1, Frankfurt-am-Main, West Germany

Copyright © 1977 John L. Dillon

All Rights Reserved. No part of this publication may be reproduced, stored in a retrieval system or transmitted in any form or by any means: electronic, electrostatic, magnetic tape, mechanical, photocopying, recording or otherwise, without permission in writing from the publishers

First edition 1968
Second edition 1977

Library of Congress Cataloging in Publication Data
Dillon, John L.
  The analysis of response in crop and livestock production.
  Includes bibliographies and indexes.
  1. Farm management—Mathematical models.
  2. Agriculture—Statistical methods. 3 Response surfaces (Statistics). 4. Fertilizers and manures—Mathematical models. 5. Feed utilization efficiency—Mathematical models. I. Title.
  S566.D5 1977   658.5   76-21327
  ISBN 0-08-021118-6
  ISBN 0-08-021115-1 pbk.

Typeset by Cotswold Typesetting Ltd and printed in Great Britain by Glevum Press, Gloucester

TO
RITA, MIKE, CAS, MATT, SUE
MARTIN, JOHN, TIM AND

Contents

Preface to Second Edition — xi

1 Response Curves and Surfaces — 1
1.1 Notation — 1
1.2 Theory of Response — 2
1.3 Variable, Fixed and Unimportant Input Factors — 4
1.4 Single Variable Input — 4
1.5 Two Variable Inputs — 8
1.6 Numerical Example with Two Variable Inputs — 18
1.7 n Variable Inputs — 23
1.8 Further Reading — 25
1.9 Exercises — 27

2 Efficiency in Response — 30
2.1 Purposes of Response Analysis — 30
2.2 Best Operating Conditions — 30
2.3 Output Gains and Input Losses — 31
2.4 Single Variable Input — 33
2.5 Two Variable Inputs — 36
2.6 n Variable Inputs — 39
2.7 Multiple Response without Input Control — 40
2.8 Multiple Response with Input Control — 42
2.9 Constraints on the Objective Function — 44
    2.9.1 Fixed-output Constraints — 44
    2.9.2 Fixed-outlay Constraints — 52
2.10 Further Reading — 59
2.11 Exercises — 61

viii   CONTENTS

3 Response Efficiency over Time — 64

- 3.1 Introduction — 64
- 3.2 Time Influences on Response — 64
- 3.3 Time-Price Effects — 66
- 3.4 Time and the Objective Function — 67
- 3.5 Planning over Time — 68
- 3.6 Unconstrained Profit Maximization over Time — 69
  - 3.6.1 Without Time Preference — 70
  - 3.6.2 With Time Preference — 73
  - 3.6.3 Numerical Example — 76
- 3.7 Constrained Profit Maximization over Time — 77
- 3.8 Time Classification of Response Processes — 78
- 3.9 Examples of Time-dependent Response Analysis — 80
  - 3.9.1 Fertilizer for Multi-harvest Crops — 81
  - 3.9.2 Feeding Period and Rations for Broilers — 83
  - 3.9.3 Livestock Production from Pasture Grazing — 86
  - 3.9.4 Crop Production with Fertilizer Carry-over — 93
- 3.10 Further Reading — 96
- 3.11 Exercises — 100

4 Response Efficiency under Risk — 102

- 4.1 Introduction — 102
- 4.2 Sources of Risk — 104
  - 4.2.1 Yield Uncertainty — 104
  - 4.2.2 Price Uncertainty — 106
- 4.3 Risk and the Objective Function — 106
  - 4.3.1 Profit Uncertainty — 106
  - 4.3.2 Expected Utility and Subjective Probability — 107
  - 4.3.3 Utility Objective Function — 109
- 4.4 Best Operating Conditions under Risk — 112
  - 4.4.1 Single Decision Variable — 113
  - 4.4.2 Multiple Decision Variables — 120
  - 4.4.3 Constrained Multiple Response — 122
  - 4.4.4 Time and Risk Together — 124
  - 4.4.5 Interrelated Yield and Price Risks — 126
  - 4.4.6 Effect of Skewness — 127

			CONTENTS	ix
4.5	Empirical Appraisal under Risk			129
	4.5.1	Specification of the Utility Function		129
	4.5.2	Evaluation of Best Operating Conditions		131
	4.5.3	Specification of the Probability Distributions		134
4.6	Stochastic Dominance Analysis			138
4.7	Further Reading			143
4.8	Exercises			146

5 Difficulties in Response Research — 149

5.1	Introduction			149
5.2	Experimental Design			150
	5.2.1	Designs for Response Surface Estimation		152
	5.2.2	Choice of Design		157
	5.2.3	Pen-feeding Trials		159
5.3	Statistical Estimation			160
	5.3.1	Least-squares Regression		160
	5.3.2	Combining Cross-section and Time-series Data		161
	5.3.3	Multi-equation and Other Models		162
	5.3.4	Series of Experiments		163
	5.3.5	Economic versus Statistical Significance		163
5.4	Response Variability over Space and Time			165
5.5	Choice of Response Model			169
	5.5.1	Linear Response and Plateau Model		170
5.6	Economics of Response Research			172
5.7	Farm versus Experimental Response			174
5.8	Making Farm Recommendations			175

REFERENCES	177
AUTHOR INDEX	197
SUBJECT INDEX	203

The Analysis of Response in Crop and Livestock Production

SECOND EDITION

by

JOHN L. DILLON

Professor of Farm Management
University of New England, Armidale, Australia

PERGAMON PRESS

OXFORD · NEW YORK · TORONTO · SYDNEY
PARIS · FRANKFURT

U. K.	Pergamon Press Ltd., Headington Hill Hall, Oxford OX3 0BW, England
U. S. A.	Pergamon Press Inc., Maxwell House, Fairview Park, Elmsford, New York 10523, U.S.A.
CANADA	Pergamon of Canada Ltd., 75 The East Mall, Toronto, Ontario, Canada
AUSTRALIA	Pergamon Press (Aust.) Pty. Ltd., 19a Boundary Street, Rushcutter Bay, N.S.W. 2011, Australia
FRANCE	Pergamon Press SARL, 24 rue des Ecoles, 75240 Paris, Cedex 05, France
WEST GERMANY	Pergamon Press GmbH, 6242 Kronberg-Taunus, Pferdstrasse 1, Frankfurt-am-Main, West Germany

Copyright © 1977 John L. Dillon

All Rights Reserved. No part of this publication may be reproduced, stored in a retrieval system or transmitted in any form or by any means: electronic, electrostatic, magnetic tape, mechanical, photocopying, recording or otherwise, without permission in writing from the publishers

First edition 1968
Second edition 1977

Library of Congress Cataloging in Publication Data
Dillon, John L.
The analysis of response in crop and livestock production.
Includes bibliographies and indexes.
1. Farm management—Mathematical models.
2. Agriculture—Statistical methods. 3 Response surfaces (Statistics). 4. Fertilizers and manures—Mathematical models. 5. Feed utilization efficiency—Mathematical models. I. Title.
S566.D5 1977   658.5   76-21327
ISBN 0-08-021118-6
ISBN 0-08-021115-1 pbk.

Typeset by Cotswold Typesetting Ltd and printed in Great Britain by Glevum Press, Gloucester

TO
RITA, MIKE, CAS, MATT, SUE, ROSIE,
MARTIN, JOHN, TIM AND DOM

Contents

PREFACE TO SECOND EDITION xi

1 Response Curves and Surfaces 1
    1.1   Notation 1
    1.2   Theory of Response 2
    1.3   Variable, Fixed and Unimportant Input Factors 4
    1.4   Single Variable Input 4
    1.5   Two Variable Inputs 8
    1.6   Numerical Example with Two Variable Inputs 18
    1.7 n Variable Inputs 23
    1.8   Further Reading 25
    1.9   Exercises 27

2 Efficiency in Response 30
    2.1   Purposes of Response Analysis 30
    2.2   Best Operating Conditions 30
    2.3   Output Gains and Input Losses 31
    2.4   Single Variable Input 33
    2.5   Two Variable Inputs 36
    2.6 n Variable Inputs 39
    2.7   Multiple Response without Input Control 40
    2.8   Multiple Response with Input Control 42
    2.9   Constraints on the Objective Function 44
            2.9.1   Fixed-output Constraints 44
            2.9.2   Fixed-outlay Constraints 52
    2.10 Further Reading 59
    2.11 Exercises 61

viii CONTENTS

3 Response Efficiency over Time — 64

- 3.1 Introduction — 64
- 3.2 Time Influences on Response — 64
- 3.3 Time-Price Effects — 66
- 3.4 Time and the Objective Function — 67
- 3.5 Planning over Time — 68
- 3.6 Unconstrained Profit Maximization over Time — 69
  - 3.6.1 Without Time Preference — 70
  - 3.6.2 With Time Preference — 73
  - 3.6.3 Numerical Example — 76
- 3.7 Constrained Profit Maximization over Time — 77
- 3.8 Time Classification of Response Processes — 78
- 3.9 Examples of Time-dependent Response Analysis — 80
  - 3.9.1 Fertilizer for Multi-harvest Crops — 81
  - 3.9.2 Feeding Period and Rations for Broilers — 83
  - 3.9.3 Livestock Production from Pasture Grazing — 86
  - 3.9.4 Crop Production with Fertilizer Carry-over — 93
- 3.10 Further Reading — 96
- 3.11 Exercises — 100

4 Response Efficiency under Risk — 102

- 4.1 Introduction — 102
- 4.2 Sources of Risk — 104
  - 4.2.1 Yield Uncertainty — 104
  - 4.2.2 Price Uncertainty — 106
- 4.3 Risk and the Objective Function — 106
  - 4.3.1 Profit Uncertainty — 106
  - 4.3.2 Expected Utility and Subjective Probability — 107
  - 4.3.3 Utility Objective Function — 109
- 4.4 Best Operating Conditions under Risk — 112
  - 4.4.1 Single Decision Variable — 113
  - 4.4.2 Multiple Decision Variables — 120
  - 4.4.3 Constrained Multiple Response — 122
  - 4.4.4 Time and Risk Together — 124
  - 4.4.5 Interrelated Yield and Price Risks — 126
  - 4.4.6 Effect of Skewness — 127

		CONTENTS	ix
4.5		Empirical Appraisal under Risk	129
	4.5.1	Specification of the Utility Function	129
	4.5.2	Evaluation of Best Operating Conditions	131
	4.5.3	Specification of the Probability Distributions	134
4.6		Stochastic Dominance Analysis	138
4.7		Further Reading	143
4.8		Exercises	146

5 Difficulties in Response Research 149

5.1		Introduction	149
5.2		Experimental Design	150
	5.2.1	Designs for Response Surface Estimation	152
	5.2.2	Choice of Design	157
	5.2.3	Pen-feeding Trials	159
5.3		Statistical Estimation	160
	5.3.1	Least-squares Regression	160
	5.3.2	Combining Cross-section and Time-series Data	161
	5.3.3	Multi-equation and Other Models	162
	5.3.4	Series of Experiments	163
	5.3.5	Economic versus Statistical Significance	163
5.4		Response Variability over Space and Time	165
5.5		Choice of Response Model	169
	5.5.1	Linear Response and Plateau Model	170
5.6		Economics of Response Research	172
5.7		Farm versus Experimental Response	174
5.8		Making Farm Recommendations	175

REFERENCES 177

AUTHOR INDEX 197

SUBJECT INDEX 203

Preface to Second Edition

DESPITE a variety of revisions and the addition of much new material, the purpose of the second edition of this little text remains unchanged. As with the first edition, its aim is to provide an introductory outline of the analytical principles involved in appraising the efficiency of crop-fertilizer and livestock-feed response.

In Chapter 1 an outline is given of the theory of crop and livestock response and of the purely physical implications of this theory. Chapters 2, 3 and 4, going beyond physical considerations, are concerned with the efficiency of response. Chapter 2 presents the principles involved in ascertaining best operating conditions for response processes in which time and risk play no role or may be reasonably ignored. These principles are extended in Chapter 3 to cover time considerations in response efficiency, and in Chapter 4 (a complete new chapter) to cover response efficiency under yield and price uncertainty. As an epilogue, Chapter 5 surveys some of the major difficulties involved in the implementation of a programme of crop or livestock response research. Variously, these problems relate to experimental design, the economics of response research, crop versus livestock experiments, statistical estimation, choice of response function form, multiple response, spatial and temporal variability in response, the relation between experimental and farm response, and farmer recommendations from response research.

As well as a variety of minor revisions, a number of major additions have been included in this second edition. These relate to the consideration of risk in determining best operating conditions, the analysis of input carry-over effects and our general discussion of difficulties in response research. As in the first edition, the temptation of attempting to cover all the refinements to response analysis discussed in the journal literature has been avoided so as to maintain the text's introductory nature. The relevant literature, however, has been extensively surveyed and referenced. Many further readings are noted at the end of Chapters 1, 2, 3 and 4 and throughout Chapter 5. Of these references, two monographs stand out as being of particular relevance—Frisch's *Theory of Production* for its thoroughgoing theoretical development of both the technical and economic aspects of production, and Heady and Dillon's *Agricultural Production Functions* for its empirical orientation to crop and livestock response analysis.

As a primer on response analysis, it is hoped the present text will help fill the gap in providing students of both agricultural science and economics with a simple but formal exposition of the why, how and wherefore of the principles of crop and livestock response analysis, thereby helping to further co-operative effort among biological and economic researchers. This is not to say that the principles enunciated here are thought of as all-important in the real world. To some extent they are no more than ideals unlikely of achievement. This must be so, given the uncertainties that exist in the real world and the fact that crop and livestock response processes are generally embedded in larger response systems—both biological and economic—whose ramifications are not irrelevant. None the less, it is hoped this introduction will provide some of the know-how needed in establishing the massive programme of crop and livestock response analysis required if countries are to obtain full benefit from their agricultural resources. The more efficient a country's agriculture is, the better fed its people can be; and the more resources there can be available to satisfy people's needs and aspirations beyond the essentials of food and fibre. Therein lies the crucial importance of the agricultural scientist. By manipulating crop and livestock response phenomena so that they better serve society's needs, he can ensure both the more efficient production of food and the release of resources needed for non-agricultural development.

For helpful comment and stimulus in preparing this second edition I am especially grateful to Jock Anderson and John Kennedy. My thanks are also due to Randy Barker, George Battese, David Boyd, Bob Cate, Tony Chisholm, Jeff Colwell, Barry Dent, Libby and Ross Drynan, Norm Dudley, Wayne Fuller, David Godden, John Pesek, Jim Ryan and Jim Seagraves for bibliographic help sent to me in the academic extremity of Brazil. For manuscript preparation, I am greatly indebted to Nell Ferris.

Fortaleza, Brazil JOHN L. DILLON

"No aphorism is more frequently repeated in connection with field trials, than that we must ask Nature few questions, or, ideally one question at a time. The writer is convinced that this view is wholly mistaken. Nature, he suggests, will best respond to a logical and carefully thought out questionnaire, indeed, if we ask her a single question, she will often refuse to answer until some other topic has been discussed."

R. A. FISHER (1926)

assumptions of analytical models

vs

real world phenomena

CHAPTER 1

Response Curves and Surfaces

1.1 Notation

Any type of physical OUTPUT requires at least two types of INPUT.† For crops and livestock, dozens of input factors are generally essential.

We will designate the ith type of input by X_i and the output (also known as YIELD, RESPONSE or PRODUCT) by Y. As required, we will also use X_i to denote the quantity of the ith input and Y to denote the quantity of output. Whether we are using X_i to refer to the input in general or to a specific quantity of the input will be apparent from the context. If Y requires m different types of input, the i subscript on X runs from 1 to m. Y and all the X's must be non-negative quantities.

Quantity of output is determined by quantity of inputs. So we can say:

$$Y \text{ depends on } X_1, X_2, X_3, \ldots, X_m;$$

or, more briefly,

$$Y = f(X_1, X_2, X_3, \ldots, X_m). \quad (1.1)$$

Equation (1.1) says Y is some unspecified mathematical function of the quantity of X's, the exact algebraic form of this RESPONSE FUNCTION (or PRODUCTION FUNCTION as it is sometimes called) being left unspecified.

Examples

(i) If Y were wheat production, the X's would be all those factors such as available soil nutrients, climate, pests, diseases, etc. which influence

† Throughout, when a technical term from response jargon is introduced for the first time, it is printed in CAPITALS with, where necessary, an immediate definition of the term.

wheat yield. While we can usually specify the more important of these factors, we could hardly list all of them.

(ii) For some particular type of product, say Y_1, the response relation might be

$$Y_1 = 2 + 3X_1 + 2X_2 - 0.01X_1^2 - 0.02X_2^2.$$

In this case, if the levels of X_1 and X_2 were 3 and 10, respectively, the level of Y_1 would be 28·91. For some other product, say Y_2, the response function might be

$$Y_2 = 186 X_1^{.2} X_2^{.3} X_3^{.1}.$$

In both of these cases, if we knew the units in which the Y's and X's were measured, we would have complete specification of the response function. Other times, we may only know the algebraic form of the response function without knowing the numerical values of the response COEFFICIENTS or PARAMETERS. For example, we may only know that

$$Y_2 = b_0 X_1^{b_1} X_2^{b_2} X_3^{b_3}$$

without knowing the value of the parameters b_0, b_1, b_2 and b_3.

1.2 Theory of Response

In general, it is impossible to list all the input factors involved in producing a particular crop or livestock product. We have to simplify. This is done by using a THEORY OF RESPONSE based on the more important input factors. As already noted, this response theory is concerned with the quantity of crop or livestock output achieved in relation to varying input quantities. It is not concerned with growth in terms of the number of members of some biological population.

Under present conditions of knowledge, the most satisfactory simple theory of crop and livestock response is that:

(i) there is a continuous smooth causal relation between the X's (inputs) and Y (output);
(ii) DIMINISHING RETURNS prevail with respect to each input factor X_i so that the *additional* output from succeeding units of X_i becomes

less and less; indeed, beyond some peak yield, additional units of X_i may have an increasingly deleterious effect on yield;

(iii) DECREASING RETURNS TO SCALE prevail so that an equal proportionate increase in all inputs results in a less than proportionate increase in output.

Assumption (i) implies that the first derivatives $\partial Y/\partial X_i$ of response equation (1.1) exist. Assumption (ii) implies $\partial Y/\partial X_i$ decreases as X_i increases, which in turn implies that the second derivatives $\partial^2 Y/\partial X_i^2$ of response equation (1.1) exist and are negative. Assumption (iii) implies that:

$$\sum (X_i/Y)(\partial Y/\partial X_i) < 1 \quad (i = 1, 2, \ldots, m). \tag{1.2}$$

The above theory of response may not always be the simplest or best to use. For response to trace elements, for example, a discontinuous theory of response might be adequate. Other times we might be happy to assume $\partial Y/\partial X_i$ constant for some X's so that these factors exhibit CONSTANT RETURNS; or to assume that a stage of constant returns precedes diminishing returns. Indeed, it is often assumed that stages of increasing, constant, and diminishing returns occur in sequence as X_i increases. However, under the usual conditions of crop and livestock production, there seems to be no strong empirical evidence for the existence of any but diminishing returns. Accordingly, we have omitted increasing and constant returns from our theory. Also, for the moment we have omitted any explicit reference to time from our theory. Compared to other input factors, time plays a much more pervasive role in response; a role which we will examine specifically in Chapter 3 via a suitably modified version of our simple theory of response.

Note ↙ model

Any theory is just a set of simplified assumptions about how reality behaves. The usefulness and test of a theory lies in its ability to predict. So long as it predicts adequately, a simple theory is to be preferred to a complex one. Since theories can be disproved but never proved, they should be continuously tested and replaced by better ones as they become available.

Example of Response Assumptions

Suppose the response function is

$$Y = aX_1^{b_1}X_2^{b_2}.$$

We have:

(i) $\partial Y/\partial X_i = b_i Y/X_i.$
(ii) $\partial^2 Y/\partial X_i^2 = b_i(b_i - 1)Y/X_i^2.$
(iii) $\sum (X_i/Y)(\partial Y/\partial X_i) = \sum b_i.$

Equation (i) indicates the response function is continuous. From equation (ii) we can see that diminishing returns to the input factor X_i will prevail so long as b_i is greater than zero and less than one. Relative to equation (1.2), equation (iii) indicates decreasing returns to scale will prevail so long as $\sum b_i$ is less than 1.

1.3 Variable, Fixed, and Unimportant Input Factors

Response function (1.1) implies all m input factors are variable. Typically, we will be concerned with the situation in which only n inputs are considered as variables, the other $m - n$ factors either being held fixed or regarded as having so little influence that we can count them as fixed. To show that $m - n$ factors are fixed or unimportant, the response function is generally written as

$$Y = f(X_1, X_2, \ldots, X_n; X_{n+1}, \ldots, X_m) \qquad (1.3)$$

or more briefly as

$$Y = f(X_1, X_2, \ldots, X_n). \qquad (1.4)$$

1.4 Single Variable Input

We have

$$Y = f(X_1) \qquad (1.5)$$

which can be depicted as a RESPONSE CURVE. Figure 1.1 gives an example. Corresponding to the theory of response, the curve is smooth, continuous and exhibits diminishing returns.

Figure 1.1 relates to some particular set or vector of quantities X_2, $X_3, \ldots, X_m$ of the non-variable inputs. If these factors were fixed at some other level, a different response curve would generally result. In

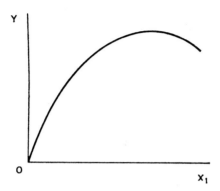

FIG. 1.1. Response curve for a single variable input.

Fig. 1.2, curve A has the non-variable factors fixed at the same level as in Fig. 1.1; curve B has them at some different (lower? higher?) level.

THEORETICAL DEDUCTIONS FROM $Y = f(X_1)$

Given the algebraic form of $Y = f(X_1)$, we can derive four quantities of interest. These are:

(i) the AVERAGE PRODUCT OF X_1, written AP_1;
(ii) the MARGINAL PRODUCT OF X_1, written MP_1;
(iii) the maximum level of Y that can be attained; and
(iv) the ELASTICITY OF RESPONSE with respect to X_1, written E_1.

The first three of these measures are physical quantities; the last, E_1, is a pure number. The common characteristic of all four quantities is that they variously relate input and output. Hence they are classified as FACTOR–PRODUCT RELATIONS (as contrasted to FACTOR–FACTOR and PRODUCT–PRODUCT RELATIONS which we will consider later).

6   THE ANALYSIS OF RESPONSE IN CROP AND LIVESTOCK PRODUCTION

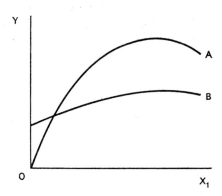

FIG. 1.2. Response curves for a single variable input with fixed factors at two different levels.

Average product of X_1 is defined as

$$AP_1 = Y/X_1. \qquad (1.6)$$

Because of diminishing returns, average product must decrease as X_1 increases since it is simply average output per unit of the variable input used.

Marginal product of X_1 is defined as

$$MP_1 = \partial Y/\partial X_1 \qquad (1.7)$$

so that it is simply the slope of the response curve. Analytically, the concept of the marginal product is much more important than that of the average product. While Y/X_1 is merely an average, $\partial Y/\partial X_1$ tells us the rate of change in Y if, at any given level of X_1, we increase X_1 by an infinitesimal amount. In other words, the marginal product tells us what happens to Y at any X_1 level as marginal changes occur in X_1. Though it is evaluated at a point on the response curve, $\partial Y/\partial X_1$ is measured in units of Y per unit of X_1 since it is a rate of change.

Note that diminishing returns implies that marginal product must always be less than average product. This is obvious from Fig. 1.1 since the slope of the response curve (i.e. MP_1) at any point is always less than the slope of a radius (i.e. AP_1) drawn from the origin to that point.

Maximum output occurs where marginal product is zero. The second-order conditions for a maximum are automatically satisfied due to the assumption of decreasing returns to scale (which is the same as diminishing returns in the case of a single variable input).† Beyond that level of X_1 which maximizes Y, $\partial Y/\partial X_1$ is increasingly negative; additional input having an increasingly deleterious or poisonous effect on output (until output is driven to zero).

Elasticity of response with respect to X_1 is defined as the relative change in Y divided by the relative change in X_1 which caused the given change in Y. Algebraically, in incremental units we have

$$E_1 = (\Delta Y/Y)/(\Delta X_1/X_1). \tag{1.8}$$

Estimated at a particular point on the response curve, equation (1.8) can be written as

$$E_1 = (\partial Y/\partial X_1)(X_1/Y) \tag{1.9}$$

$$= MP_1/AP_1. \tag{1.10}$$

Since MP_1 and AP_1 are measured in similar units, E_1 is a pure number. It is most conveniently interpreted as the percentage change in Y resulting from a 1 per cent increase in X_1. Since our theory of response implies MP_1 must always be less than AP_1, we can see from equation (1.10) that E_1 must always be less than 1 to be consistent with the theory both in terms of diminishing returns to the factor and in terms of decreasing returns to scale.

To sum up, we have considered the single-variable response process as a continuous response function within the framework of a well-defined and (so far as statistical considerations allow) empirically refutable theory of response. Scientifically, such an approach is obviously much more attractive and rewarding than the mere description of single-variable response by plotting or tabling a few discrete points in an input–output diagram or table. Even so, the single variable input response function has two disadvantages. For a start, it tells us nothing of the relation between X_1 and the other Xs' when they vary. As well, it tells us naught of the relation between Y and these other input factors. Relative

† Throughout, we will assume the necessary second-order conditions are satisfied. Their derivation is detailed in Frisch (1965, chs. 5–8 *passim*).

to Fig. 1.2, $Y = f(X_1)$ gives us no information on the move from curve A to curve B or vice versa. Nor does it tell us how AP_1, MP_1, and E_1 may be influenced by changes in the factors $X_2, X_3, \ldots, X_m$. For such deductions we have to study response functions involving more than one variable input.

Example of $Y = f(X_1)$

Suppose $Y = f(X_1)$ is specified as

$$Y = 10 + 100X_1 - X_1^2$$

with Y measured in quintals per hectare and X_1 in kilograms per hectare. Applying the formulae of equations (1.6), (1.7), and (1.9), we have

$$AP_1 = 10/X_1 + 100 - X_1$$

$$MP_1 = 100 - 2X_1$$

$$E_1 = (100 - 2X_1)/(10/X_1 + 100 - X_1).$$

Decreasing returns prevail because $d(MP_1)/dX_1$ is negative. It equals minus 2 quintals of Y per kilogram of X_1. Substitution of particular X_1 values into the above equations gives, respectively, predictions of Y, AP_1, MP_1, and E_1 at these X_1 levels. Some such predictions of these factor–product relationships are given in Table 1.1. Setting the marginal product equation equal to zero indicates maximum output will occur when X_1 equals 50 kilograms, the corresponding output being 2510 quintals.

1.5 Two Variable Inputs

For the two-variable response function, we have

$$Y = f(X_1, X_2). \tag{1.11}$$

Unlike the single-variable situation, this two-variable input function cannot be represented by a single curve. It depicts, not a curve, but a surface in the three-dimensional space with axes X_1, X_2 and Y. Two diagrammatic possibilities exist. Either we can show the response

TABLE 1.1

AVERAGE PRODUCT, MARGINAL PRODUCT AND ELASTICITY OF RESPONSE OF X_1, AND PREDICTED OUTPUT FOR VARIOUS X_1 LEVELS IN $Y = 10 + 100X_1 - X_1^2$

X_1 (kilograms)	AP_1 (quintals per kilogram)	MP_1 (quintals per kilogram)	E_1	Y (quintals)
0	n.d.	100	n.d.	10
10	91·00	80	0·88	910
20	80·50	60	0·75	1610
30	70·33	40	0·57	2110
40	60·25	20	0·33	2410
50	50·20	0	0·00	2510
60	40·17	−20	−0·50	2410

n.d.: not defined.

function as a surface in three dimensions, or we can depict it by a series of curves in two dimensions. The latter is to be preferred for analytic purposes.

Figure 1.3 exemplifies the three-dimensional alternative.

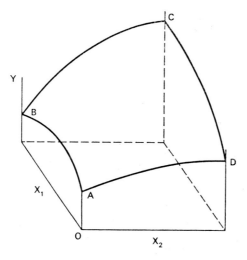

FIG 1.3. Response surface for two variable inputs.

10 THE ANALYSIS OF RESPONSE IN CROP AND LIVESTOCK PRODUCTION

The RESPONSE SURFACE ABCD is merely the surface of the "output-hill" traced out as X_1 and X_2 increase. Changes in the specification of the response function would cause changes in the shape and location of the surface.

Figure 1.4 shows the preferred alternative of depicting $Y = f(X_1, X_2)$ by a series of output contours to give a bird's-eye view of the surface. These yield contours are known as ISOQUANTS. Each isoquant is the locus of all combinations of X_1 and X_2 that produce the same level of output.

The isoquants of Fig. 1.4 are equally spaced in terms of Y values. In consequence, as implied by the theory of response (why?), the distance between isoquants measured in terms of X_1 or X_2 becomes larger as Y increases to its maximum.

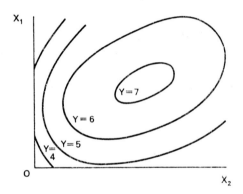

FIG. 1.4. Isoquants for two variable inputs.

Just as Fig. 1.4 represents $Y = f(X_1, X_2)$ by a series of curves in the (X_1, X_2)-plane, the response function could be represented by a series of single-variable response curves in the (Y, X_1) or (Y, X_2)-planes. For example, Fig. 1.2 might be interpreted as representing $Y = f(X_1, X_2)$ with X_2 fixed at two levels A and B.

Analogous to the single variable input case, Figs. 1.3 and 1.4 relate to some particular set of levels for the fixed or unimportant factors. If the levels of these fixed factors were changed, the shape of the surface would also be changed.

THEORETICAL DEDUCTIONS FROM $Y = f(X_1, X_2)$

If we know the algebraic form of $Y = f(X_1, X_2)$ we can easily derive all the information available from the single-variable functions $Y = f(X_1)$ and $Y = f(X_2)$. To obtain these single variable functions we simply take X_2 or X_1, respectively, as fixed at some level. For example, if the two-variable response function is

$$Y = a(1 - 10^{-c_1 X_1} - 10^{-c_2 X_2}),$$

the single-variable function in X_1 with X_2 fixed at some particular level is

$$Y = a(b - 10^{-c_1 X_1})$$

where b is equal to $(1 - 10^{-c_2 X_2})$.

The only qualitative change that occurs in factor–product relations when there is more than one variable input is that diminishing returns (which relates to a particular variable factor) is no longer equivalent to decreasing returns to scale (which relates to all variable factors). Thus the response function

$$Y = X_1^{\cdot 3} X_2^{\cdot 8}$$

exhibits diminishing returns to X_i but not decreasing returns to scale. As equation (1.2) shows, the type and extent of returns to scale that prevail at a given point on the response function are given by the sum of the individual elasticities of response of the variable factors at that point. In short, returns to scale is the *overall* elasticity of response to the variable factors.

According as this overall elasticity is greater than, equal to, or less than 1, we have increasing, constant, or decreasing returns to scale. As already noted, however, we assume crop and livestock response exhibits only decreasing returns.

Over and above the factor–product relations, with two variable inputs we enter the realm of FACTOR–FACTOR RELATIONS. These consist of:

(i) the family of isoquant equations;
(ii) the RATE OF TECHNICAL SUBSTITUTION of X_i for X_j, written RTS_{ij};
(iii) the ELASTICITY OF SUBSTITUTION of X_i for X_j, written ES_{ij};
(iv) the family of ISOCLINE equations; and
(v) the RIDGE-LINE equations.

Isoquant equations, being *loci* of input combinations that yield a fixed level of output, are obtained by rearranging the response function to give one input as a function of the other with Y regarded as fixed. Thus if Y^* denotes a fixed level of Y, the isoquant function for output Y^* is

$$X_1 = f(X_2; Y^*). \qquad (1.12)$$

For various levels of Y, this function gives a family of isoquant equations. Obviously the shape and location of the isoquants depend on the form of the parent response function.

The rate of technical substitution of X_1 for X_2 is given by the isoquant slope. At any point on an isoquant we have

$$RTS_{12} = \partial X_1/\partial X_2 \qquad (Y = Y^*) \qquad (1.13)$$
$$= 1/RTS_{21}. \qquad (1.14)$$

The rate of technical substitution of X_1 for X_2 tells us the rate at which we have to substitute X_1 for X_2 if we decrease X_2 by an infinitesimal amount and wish to maintain output unchanged.† It can range from minus to plus infinity. Being a rate, RTS_{12} is measured in units of X_1 per unit of X_2.

Elasticity of substitution of X_1 for X_2 is defined as the relative change in X_1 divided by the relative change in X_2 if we substitute X_1 for X_2 while keeping output unchanged.‡ We thus have

$$ES_{12} = (\Delta X_1/X_1)/(\Delta X_2/X_2) \qquad (Y = Y^*) \qquad (1.15)$$

which, estimated at a particular point on the isoquant, is

$$ES_{12} = (\partial X_1/\partial X_2)(X_2/X_1) \qquad (1.16)$$
$$= (RTS_{12})(X_2/X_1) \qquad (1.17)$$
$$= 1/ES_{21}. \qquad (1.18)$$

† Though it is a rate and *not* a marginal rate, RTS_{12} is often referred to as the "marginal rate of (technical) substitution of X_1 for X_2" and denoted MRS_{12} or $MRTS_{12}$. Too, it is sometimes defined as the negative of the isoquant slope. Compare, e.g. Bilas (1967, pp. 105–9), Heady (1952, pp. 140–4), Henderson and Quandt (1971, pp. 59–60) and Leftwich (1970, pp. 134–6). See also Newman (1965, p. 48).

‡ This definition, usual in response analysis, differs from that used by economists in studying factor substitution. Their definition is of ES_{ij} as the proportional rate of change in RTS_{ij}. Compare Allen (1938, p. 341) and Heady (1952, p. 145).

Being an elasticity, ES_{ij} is a pure number. It can range from minus to plus infinity; and is conveniently interpreted as the percentage change in X_i needed to maintain Y unchanged if we change X_j by 1 per cent.

Isoclines are defined as the *loci* of all combinations of X_1 and X_2 which have the same rate of technical substitution. Hence they constitute paths up or down the response surface joining points of equal curvature on the isoquants.

The family of isocline equations is derived by solving

$$\partial X_1/\partial X_2 = k \qquad (1.19)$$

to obtain X_1 as a function of X_2; k being the value of RTS_{12} which specifies a particular isocline. For example, in Fig. 1.4 the isocline for a RTS_{12} of -0.6 could be drawn by connecting up the points at which the isoquants have a slope of -0.6.

From equations (1.12) and (1.19) it is obvious that the shape and location of the isoclines depend upon the form of the parent response function. In particular, if this exhibits a maximum, the isoclines will pass through it. The response function also determines whether or not the isoclines emanate from the origin.

Ridge-lines are those two special isoclines for which RTS_{12} is equal to zero or infinity, as implied respectively by their equations:

$$\partial X_1/\partial X_2 = 0, \qquad (1.20)$$

$$\partial X_2/\partial X_1 = 0. \qquad (1.21)$$

So long as the response function exhibits a specific maximum, the ridge-lines divide the surface into four sectors. As discussed later, these are of varying significance due to the fact that the ridge-lines mark the boundary between rational and irrational combinations of inputs.

The rate of technical substitution is by far the most basic and important of the above factor–factor concepts. Accordingly, we will examine this concept in more detail before presenting a numerical example for the case of two variable inputs.

RATE OF TECHNICAL SUBSTITUTION

Consider the isoquant segment AB in Fig. 1.5. On AB, all combinations of X_1 and X_2 yield the same output. So, as we move from one point

14   THE ANALYSIS OF RESPONSE IN CROP AND LIVESTOCK PRODUCTION

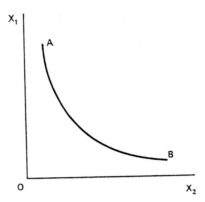

FIG. 1.5. Isoquant segment with diminishing negative rate of technical substitution.

to another on the isoquant, we can talk of X_1 substituting for X_2 or vice versa. This is not to say that X_1 necessarily replaces X_2 directly in any basic chemical or physiological process in the plant or animal. It merely reflects the fact that within limits (as defined by the location of the isoquant) the same yield can result from various combinations of quite diverse nutrients. For example, within limits, nitrate fertilizer can substitute for irrigation water (and vice versa) in wheat production; just as protein and carbohydrate, again within limits, can substitute for one another in broiler production. To put it another way, a smooth continuous response function $Y = f(X_1, X_2)$ implies a continuous smooth surface. In turn, this necessarily implies yield contours and hence substitution possibilities.

Moving up the isoquant segment AB of Fig. 1.5, the slope of the isoquant, being negative, diminishes. Accordingly, since $\partial X_1/\partial X_2$ is defined as RTS_{12}, on AB we are said to have a diminishing negative rate of technical substitution of X_1 for X_2.†

By similar reasoning, for a complete move around an individual isoquant we can distinguish four types of rates of technical substitution

† As we move from B to A it becomes increasingly difficult to substitute X_1 for X_2, a larger and larger amount of X_1 being required to replace a unit of X_2. Therefore, on AB, RTS_{12} is sometimes said to be an increasing rate of technical substitution. See Allen (1938, p. 341).

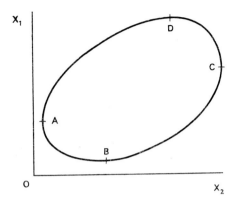

FIG. 1.6. Isoquant segments with different types of technical rates of substitution.

between two inputs. For example, consider the isoquant ABCD of Fig. 1.6, the points A, B, C, and D being where the ridge-lines cross the isoquant. We have that:

from C to B, RTS_{12} is diminishing and positive;

from B to A, RTS_{12} is diminishing and negative;

from C to D, RTS_{12} is increasing and negative;

from D to A, RTS_{12} is increasing and positive.

Corresponding to Fig. 1.6, if we plot RTS_{12} against X_2 for a move around a particular isoquant, we obtain graphs of the form shown in Fig. 1.7. These illustrate the movement of RTS_{12} between minus and plus infinity as we circle an isoquant.

The rate of technical substitution also provides the bridge between factor–product and factor–factor relations. As shown below, it is equal to the negative inverse ratio of the marginal products. First, however, consider Fig. 1.8 which shows an isoquant and ridge-lines. R is the peak of the surface where MP_1 and MP_2 are both zero. Along the ridge-line AC, MP_1 is zero since for a given level of X_2, the corresponding point on AC locates the highest isoquant or highest yield that can be achieved by

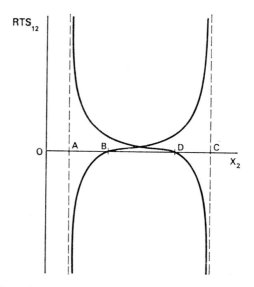

Fig. 1.7. RTS_{12} around an isoquant plotted against X_2.

varying X_1 with X_2 fixed at that level. Above AC, for given X_2, yield declines as X_1 increases. Thus, for all combinations of X_1 and X_2 that lie above the ridge-line AC, $\partial Y/\partial X_1$ (i.e. MP_1) is negative; below AC, it is always positive. Similarly, $\partial Y/\partial X_2$ (i.e. MP_2) is zero along ridge-line BD, positive to the left, and negative to the right.

These relationships are important for response efficiency, as discussed in Chapter 2. They are also relevant to calculating RTS_{ij}. Being typically elliptical, the isoquant function is algebraically complex compared to its parent response function. Rather than differentiate the multi-valued isoquant equation directly, it is much easier to calculate technical substitution rates by using the following relationships†:

† The relationships can be deduced algebraically by total implicit differentiation of $Y^* = f(X_1, X_2)$ with respect to X_2. Thus, since $X_1 = f(X_2; Y^*)$, we have
$$(\partial Y/\partial X_1)(dX_1/dX_2) + \partial Y/\partial X_2 = 0,$$
so that the slope of the isoquant, dX_1/dX_2 (i.e. RTS_{12}), is equal to: $-(\partial Y/\partial X_2)/(\partial Y/\partial X_1)$ or $-MP_2/MP_1$.

RESPONSE CURVES AND SURFACES   17

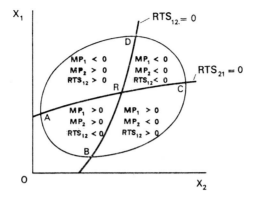

Fig. 1.8. Relation between marginal products and ridge-line locations.

$$RTS_{12} = \partial X_1/\partial X_2 \quad (Y = Y^*) \quad (1.22)$$

$$= -(\partial Y/\partial X_2)/(\partial Y/\partial X_1) \quad (1.23)$$

$$= -(MP_2/MP_1) \quad (1.24)$$

$$= 1/RTS_{21}. \quad (1.25)$$

In using this short-cut procedure, a negative sign must be placed on the inverse ratio of the marginal products—as is indicated by equation (1.24). The reason for this negative sign arises from the marginal product considerations illustrated in Fig. 1.8. For example, if we want RTS_{12} for a point in the region OARB in which *negative* rates of technical substitution prevail, we must attach a negative sign to the ratio of the *positive* marginal products. Likewise, in the other three regions marked off by the ridge-lines, a negative sign must be attached to the inverse ratio of the marginal products to achieve consistency in sign between this ratio and the rate of technical substitution.

The same result can be reached intuitively by another route. For example, in the region OARB, when we move along the isoquant AB by an infinitesimal increase in X_2, our rate of *gain* in output per unit of X_2 is MP_2. Our rate of *loss* in output per unit of X_1 is MP_1. Hence, in the region OARB, use of equation (1.24) implies attaching a plus sign to MP_2 since it is a gain, and a minus sign to MP_1 since it is a loss. An

18 THE ANALYSIS OF RESPONSE IN CROP AND LIVESTOCK PRODUCTION

analogous intuitive argument is applicable to each of the other regions of Fig. 1.8.

Relative to Fig. 1.8, note that the theory of response does not imply that the origin must lie within the region of positive marginal products. It depends on the algebraic form of the response function. A configuration such as that in Fig. 1.9 is quite reasonable.

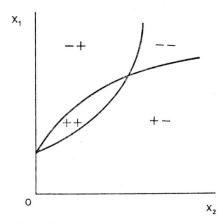

Fig. 1.9. A possible configuration of positive and negative marginal products of X_1 and X_2.

1.6 Numerical Example with Two Variable Inputs

As an illustrative example we use a wheat-fertilizer response function estimated from a field trial conducted at El Llano in the Central Valley of Chile in 1962–3.

The response function is

$$Y = 18 \cdot 846512 + 7 \cdot 586167N + 2 \cdot 469969P$$
$$- 0 \cdot 655713N^2 - 0 \cdot 397513P^2 + 0 \cdot 211423NP, \qquad (1.26)$$

where Y is quintals of wheat per hectare, N is sodium nitrate in units of 150 kilograms per hectare, and P is triple superphosphate in units of 100 kilograms per hectare.

For N the experimental range was from 0 to 7 units, and for P from 0 to 5 units. Major fixed factors were the soil type, irrigation management and the climate of 1962-3.

Predicted yields over the experimental range of N and P are given in Table 1.2. Each row gives predicted yields for the single-variable response function $Y = f(N)$ with superphosphate fixed at the level specified for that row. Likewise each column corresponds to some $Y = f(P)$ curve. Algebraically, for example, $Y = f(P)$ for N of 2·0 is obtained from equation (1.26) by setting N equal to 2·0. This gives the single-variable response function:

$$Y = 31 \cdot 395994 + 2 \cdot 892815P - 0 \cdot 397513P^2. \qquad (1.27)$$

TABLE 1.2

QUINTALS OF WHEAT PER HECTARE PREDICTED FOR SPECIFIED COMBINATIONS OF APPLIED PHOSPHATE AND NITRATE FERTILIZER, EL LLANO, CENTRAL VALLEY, CHILE, 1962-3[a]

Kilograms of superphosphate applied per hectare	Kilograms of sodium nitrate applied per hectare						
	0	150	300	450	600	750	900	1050
	quintals per hectare						
0	18·8	25·8	31·4	35·7	38·7	40·4	40·8	39·8
100	20·9	28·3	33·9	38·4	41·6	43·5	44·1	43·4
200	22·0	29·5	35·6	40·3	43·7	45·8	46·6	46·1
300	22·7	30·2	36·5	41·4	45·1	47·4	48·4	48·1
400	22·4	30·1	36·6	41·8	45·6	48·1	49·3	49·3
500	21·3	29·2	35·9	41·3	45·3	48·1	49·5	49·6

[a] Based on the response function of equation (1.26).

Figure 1.10 shows the three-dimensional response surface of response function (1.26). In Fig. 1.10(a) the surface is shown with interlaced single-variable response curves. Isoquants around the surface are shown in Fig. 1.10(b). A bird's-eye view of the isoquants is given in Fig. 1.11, the area within the dotted lines being the experimental range of N and P.

The isoquant function $P = f(N; Y^*)$ is obtained by substituting the required fixed level of output for Y in equation (1.26) and rearranging to obtain the equation:

$$P = 3 \cdot 10678 + 0 \cdot 26593 N \pm (57 \cdot 06287 + 20 \cdot 73648 N$$
$$- 1 \cdot 57882 N^2 - 2 \cdot 51564 Y*)^{\frac{1}{2}}. \quad (1.28)$$

Marginal products per unit of P and N are obtained directly as the first derivatives of response function (1.26). We have:

$$MP_P = 2 \cdot 469969 - 0 \cdot 795026 P + 0 \cdot 211423 N, \quad (1.29)$$
$$MP_N = 7 \cdot 586167 - 1 \cdot 311426 N + 0 \cdot 211423 P. \quad (1.30)$$

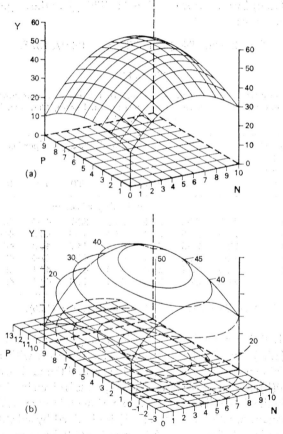

Fig. 1.10. Three-dimensional response surface of response function (1.26): (a) with interlaced single-variable response curves and (b) with yield isoquants.

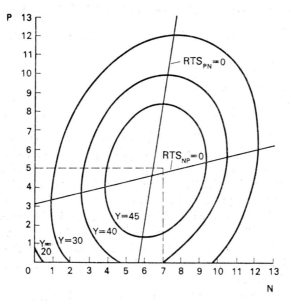

Fig. 1.11. Isoquant diagram of response function (1.26) with ridge-lines and experimental range.

Notice that because of the form of the parent response function, MP_P and MP_N depend upon the level of both variables. Both marginal products imply diminishing returns.

Rates of technical substitution are obtained as the negatively-signed inverse ratio of the marginal products. We have:

$$RTS_{PN} = -(7 \cdot 586167 - 1 \cdot 311426N + 0 \cdot 211423P)/$$

$$(2 \cdot 469969 - 0 \cdot 795026P + 0 \cdot 211423N) \quad (1.31)$$

$$= 1/RTS_{NP}. \quad (1.32)$$

Elasticities of substitution are derived as in equation (1.17) for any surface point (N, P) as the value of the product $RTS_{PN}(N/P)$ or $RTS_{NP}(P/N)$. Thus we have:

22 THE ANALYSIS OF RESPONSE IN CROP AND LIVESTOCK PRODUCTION

$$ES_{PN} = -(7 \cdot 586167 P^{-1} - 1 \cdot 311426 NP^{-1} + 0 \cdot 211423)/$$
$$(2 \cdot 469969 N^{-1} - 0 \cdot 795026 PN^{-1} + 0 \cdot 211423)$$
$$= 1/ES_{NP}. \qquad (1.33)$$

As with the marginal products, because of the quadratic form of the response function, both the rates of technical substitution and the elasticities of substitution depend upon the level of both N and P.

Predictions of MP_N, MP_P, RTS_{PN} and ES_{PN} for a few combinations of phosphate and nitrate on the 40-quintal isoquant are given in Table 1.3. The data of the table relate to kilograms of input, not the N and P units of response function (1.26). For the specified input levels, Table 1.3 also lists the prevailing returns to scale or overall elasticity of production derived as the sum of the individual response elasticities as per equation (1.2). The positive nature of the predicted marginal products indicates that the chosen fertilizer combinations all lie within the region of diminishing negative rates of technical substitution, i.e. within region OARB of Fig. 1.8.

The isocline equations are derived by setting the expression for

TABLE 1.3

MARGINAL PRODUCTS, RATES OF TECHNICAL SUBSTITUTION, ELASTICITIES OF SUBSTITUTION AND RETURNS TO SCALE AT VARIOUS COMBINATIONS OF NITRATE AND PHOSPHATE FERTILIZER ON THE 40 QUINTAL YIELD ISOQUANT FOR WHEAT AT EL LLANO, CENTRAL VALLEY, CHILE, 1962-3[a]

Nitrate (kilograms per hectare)	Phosphate (kilograms per hectare)	MP_N $(\partial Y/\partial N)$ (quintals per kilogram)	MP_P $(\partial P/\partial N)$ (quintals per kilogram)	RTS_{PN} $(\partial P/\partial N)$ (kilograms of P per kilogram of N)	ES_{PN} $(\partial P/\partial N)/(N/P)$	Returns to scale $(E_P + E_N)$
450	179·86	0·0269	0·0171	−1·572	−3·932	0·379
525	92·40	0·0213	0·0249	−0·853	−4·848	0·337
600	41·26	0·0162	0·0299	−0·540	−7·815	0·274
675	8·68	0·0113	0·0335	−0·338	−26·323	0·198

[a] Based on the response function of equation (1.26).

RTS_{PN} of equation (1.31) equal to the desired isoquant slope, say k, and solving to give P in terms of N or vice-versa. Thus we obtain

$$P = (7 \cdot 586167 + 2 \cdot 469969k - 1 \cdot 311426N + 0 \cdot 211423kN)/ \\ (0 \cdot 795026k - 0 \cdot 211423). \quad (1.34)$$

Ridge-line equations are obtained by setting the expressions for RTS_{PN} and RTS_{NP} respectively equal to zero and solving to give P in terms of N or vice versa. For RTS_{PN} equal to zero, we obtain

$$P = -35 \cdot 881465 + 6 \cdot 202854N; \quad (1.35)$$

and for RTS_{NP} equal to zero (i.e. RTS_{PN} equal to infinity) we obtain

$$P = 3 \cdot 106776 + 0 \cdot 265932N. \quad (1.36)$$

These two ridge-lines are shown in Fig. 1.11. Because of the nature of the parent response function, the ridge-lines—and all the other isoclines—are straight lines.

1.7 n Variable Inputs

For n larger than two, the advantage of the n-variable response function

$$Y = f(X_1, X_2, \ldots, X_n) \quad (1.37)$$

is that as well as giving additional information about Y, it gives information on factor–factor relations for more than one pair of factors. This extra information, however, is only obtained at some cost. The size of n will usually be determined by the size of these costs relative to the importance of understanding the role of $X_1, X_2, \ldots, X_n$ in determining Y. The n variable inputs imply a response surface in $(n + 1)$-dimensional space. For n greater than 2, description and analysis of the response process has to be in terms of algebra. The algebra involved is merely an extension of that for two variable inputs. For i or $j = 1, 2, \ldots, n$, we have the following relationships:

$$AP_i = Y/X_i, \quad (1.38)$$

$$MP_i = \partial Y/\partial X_i, \quad (1.39)$$

$$RTS_{ij} = -(MP_j/MP_i), \quad (1.40)$$

$$ES_{ij} = RTS_{ij}(X_j/X_i). \quad (1.41)$$

The isoquant surface for a particular level of output is specified by the function:

$$X_1 = f(X_2, X_3, \ldots, X_n; Y^*). \tag{1.42}$$

Isoclines joining all combinations of X_i and X_j which have the same rate of technical substitution, say k_{ij}, are given by:

$$RTS_{ij} = k_{ij}. \tag{1.43}$$

Ridge-lines or surfaces on which RTS_{ij} equals zero or infinity are derived by solving the isocline equation for k_{ij} values of zero and infinity, respectively.

Diminishing returns to X_i implies

$$\partial(MP_i)/\partial X_i < 0. \tag{1.44}$$

At any point on the response surface, the overall elasticity of response—which must be less than one for decreasing returns to scale—is given by the sum $\sum(\partial Y/\partial X_i)(X_i/Y)$.

Depending on the algebraic form of the parent response function, all of the above relations may involve up to n input variables. That is, for a response surface in $(n+1)$-dimensional space, the factor–factor and factor–product relations may correspond to surfaces in up to n or $(n+1)$-dimensional space respectively.

Example for $Y = f(X_1, X_2, \ldots, X_n)$

Suppose the n-variable input response function is specified as the power or Cobb-Douglas function

$$Y = b_0 X_1^{b_1} X_2^{b_2} \ldots X_n^{b_n} \tag{1.45}$$

$$= b_0 \Pi X_i^{b_i}. \tag{1.46}$$

For average and marginal products, we have, respectively,

$$AP_i = Y/X_i, \tag{1.47}$$

$$MP_i = b_i Y/X_i. \tag{1.48}$$

For diminishing returns to X_i, $\partial(MP_i)/\partial X_i$ must be negative. This implies b_i must be greater than zero and less than 1. Accordingly, with diminishing returns to X_i, its marginal product must always be nonnegative and decreasing. The isoquant surface is

$$X_1 = (Y^*/b_0 X_2^{b_2} X_3^{b_3} \ldots X_n^{b_n})^{1/b_1}. \tag{1.49}$$

The rate of technical substitution and elasticity of substitution between X_i and X_j are, respectively,

$$RTS_{ij} = -b_j X_i/b_i X_j, \tag{1.50}$$

$$ES_{ij} = -b_j/b_i. \tag{1.51}$$

Applying equation (1.43), the isocline equations are

$$X_i = -k_{ij}(b_i/b_j)X_j. \tag{1.52}$$

Ridge-lines for RTS_{ij} and RTS_{ji} respectively equal to zero are

$$X_i = 0, \tag{1.53}$$

$$X_j = 0. \tag{1.54}$$

The elasticity of response to X_i is

$$E_i = b_i, \tag{1.55}$$

so that, from equation (1.2), the condition for decreasing returns to scale is that the sum $\sum b_i$ must be less than one.

Obviously the power function is a rather special response function. Because of its multiplicative nature, if X_i is zero, output must also be zero. It exhibits no finite maximum, as indicated by the fact that MP_i is always positive. Correspondingly, its ridge-lines (like all its isoclines) emanate from the origin and do not converge. Indeed, the ridge-lines are the input axes; and diminishing negative rates of technical substitution between factors prevail over the entire surface. Moreover, the elasticity of response for X_i and the overall elasticity of response are constant—they do not vary across the response surface as the inputs vary.

1.8 Further Reading

The historical development of the theory of crop and livestock response has been outlined by Heady (1964), Heady and Dillon (1961, ch. 1),

Jonsson (1974), Mason (1956), Natl. Res. Council (1961), Pesek (1973) and Ryan (1972).† These references give extensive citations to the original contributions of such workers as Balmukand, Baule, Bondorff, Heady, Jensen, Liebig, Mitscherlich, Spillman, and others. Here it suffices specifically to mention only Spillman (1933), Spillman and Lang (1924), Brody (1945) and—since they appear to elsewhere have been overlooked as pioneering works—the studies of Clyde et al. (1923), Prescott (1928), Richardson and Fricke (1931) and Southee (1925).

Some discussion of the logic of response theory, mainly oriented to applications in industrial chemical processes, is to be found in Box (1954), Chew (1958), Dano (1966) and Kennard (1963). Direct discussion of the theory of crop and livestock response processes has been given by Heady (1952, chs. 2, 3, 5), Heady and Dillon (1961, ch. 2), Munson and Doll (1959) and, with a more basic biological emphasis, by Blaxter (1962b and 1964), Boyd (1972), Fried and Broeshart (1967), Helyar and Godden (1976), Sandland and Jones (1975), and Russell (1972). Many empirical examples of crop and livestock response analysis are to be found in Heady and Dillon (1961, chs. 8–15). This reference also contains a discussion (ch.6) of a more elaborate theory of response based on systems of simultaneously determined response relations rather than a single causal equation.

Various algebraic forms of response function and their implications are outlined by Heady and Dillon (1961, ch. 3), Anderson and Nelson (1975), Halter, Carter and Hocking (1957), Jonsson (1974), Nelder (1966), Swanson (1963) and, for generalizations of the power form, by de Janvry (1972a and b), and Ulveling and Fletcher (1970). See also Zellner and Revankar (1969) and Upton and Dalton (1976).

Formal treatment of the mathematics involved in response theory is to be found in Allen (1938) and Tintner (1970). Frisch (1965, chs. 5, 6, 7, 8) gives a most thoroughgoing mathematical analysis of the physical implications of response theory. As well as for his explicit development of necessary second-order conditions, Frisch's treatment is of interest for its particular emphasis on interpretations via the elasticity of response. Extensions of Frisch's analysis in an industrial context are presented by Dano (1966).

† References are listed alphabetically at the end of the volume.

1.9 Exercises

1.9.1. Accurately graph the 35-quintal isoquant for response function (1.26) and duplicate Table 1.3 for some points on this isoquant.

1.9.2. Some alternative estimates of response function (1.26) are as follows:

(i) $Y = 1.34 + 9.77N^{.5} + 2.63P^{.5} - 0.43N - 0.81P + 1.34N^{.5}P^{.5}$
(ii) $Y = 18.01 + 10.43N^{.7} + 2.98P^{.7} - 1.19N^{1.4} - 0.74P^{1.4} + 0.65N^{.7}P^{.7}$
(iii) $Y = 18.16 + 10.93N + 3.95P - 2.9N^{1.5} - 1.53P^{1.5} + 0.21NP.$

Derive the isoquant and ridge-line equations for each of these alternative estimates; and graphically compare their ridge-lines with those of response function (1.26). For each estimated response function, also graphically compare the single variable input function $Y = f(N; P = 3)$.

1.9.3. The response functions listed below illustrate some of the algebraic forms that have been used in crop and livestock response research. For each function, derive the conditions under which:

(a) diminishing returns prevail to X_i;
(b) decreasing returns to scale prevail;
(c) the marginal product of X_i is negative.

Where applicable, also derive:

(d) an isoquant equation;
(e) the ridge-line equations; and
(f) the elasticity of substitution between X_i and X_j.

Finally, draw up a table classifying the functions in terms of whether or not their implied factor–product and factor–factor relations depend on the level of all inputs, whether or not they admit negative marginal products, and whether or not their isoclines emanate from the origin.

(i) First degree polynomial in $X_i^{b_i}$:

$$Y = a_0 + a_1 X_1 + a_2 X_2,$$
$$Y = a_0 + a_1 X_1^{\frac{1}{2}} + a_2 X_2^{\frac{1}{2}}.$$

(ii) Second degree polynomial in $X_i^{b_i}$:

$$Y = a_0 + a_1 X_1 + a_{11} X_1^2,$$

$$Y^{-1} = a_0 + a_1 X_1 + a_2 X_2^{-1} + a_{11} X_1^2 + a_{22} X_2^{-2}$$
$$+ a_{12} X_1 X_2^{-1},$$

$$Y = a_0 + \sum a_i X_i^{b_i} + \sum a_{ii} X_i^{2b_i} + \sum\sum a_{ij} X_i^{b_i} X_j^{b_j}, \ (i < j),$$

$$Y = a_0 + a_1 X_1^{\frac{1}{2}} + a_2 X_2^{\frac{1}{2}} + a_{11} X_1 + a_{22} X_2^{\frac{1}{2}} + a_{12} X_1^{\frac{1}{2}} X_2^{\frac{1}{2}}.$$

(iii) Mitscherlich or Spillman function:

$$Y = M - AR^{X_1},$$

$$Y = M - \sum A_i R_i^{X_i},$$

$$Y = a_0(1 - a_1^{X_1}),$$

$$Y = a_0(1 - a_1^{X_1}) a_2^{X_1^2},$$

$$Y = a_0 \Pi (1 - a_i^{X_i + b_i}).$$

(iv) Resistance function:

$$Y^{-1} = a_0 + \sum a_i (b_i + X_i)^{-1}.$$

(v) Cobb–Douglas or power function:

$$\log Y = \log a_0 + \sum a_i \log X_i,$$

$$Y = a_0 \Pi (a_i + X_i)^{b_i}.$$

(vi) Transcendental function:

$$Y = a \Pi X_i^{b_i} e^{c_i X_i}.$$

(vii) Generalized power function:

$$Y = a \Pi X_i^{f_i(X_1, X_2, \ldots, X_n)} e^{g(X_1, X_2, \ldots, X_n)},$$

where $f_i(.)$ and $g(.)$ are polynomials of any degree.

For example,

$$Y = a X_1^{b_1 + b_2 X_2} X_2^{b_3} X_3^{b_4 + b_5 X_1 + b_6 X_1^2} e^{b_7 X_2 + b_8 X_3^2},$$

$$Y = a X_1^{b_1} X_2^{b_2} e^{b_3 + b_4 X_1 + b_5 X_2},$$

$$Y = a X_1^{b_1} X_2^{b_2}.$$

1.9.4. Interpreted as response functions, what special characteristics are exhibited by the following algebraic forms?†

(i) $Y = (2kX_1X_2 - a_1X_1^2 - a_2X_2^2)/(a_3X_1 + a_4X_2)$,

(ii) $Y = (\sum a_i X_i^{-k})^{-1/k}$,

(iii) $Y = \sum a_i X_i + b\Pi X_i^{c_i}$,

(iv) $Y = X_1^{a_1} X_2^{a_2} - X_1^{b_1} X_2^{b_2}$,

(v) $Y = (aX_1^a + \sum b_i X_1^{\beta_i} X_2^{a-\beta_i} + cX_2^a)^{1/a}$.

(vi) $Y = \begin{cases} a + bX_i & \text{for } 0 \leq X_i \leq X_i^* \\ a + bX_i^* & \text{for } X_i \geq X_i^*. \end{cases}$

† Variously, these functional forms are discussed in Allen (1938, p. 349), Arrow *et al.* (1961), Sato (1964), Upton and Dalton (1976), and Van Moeseke (1965).

CHAPTER 2

Efficiency in Response

2.1 Purposes of Response Analysis

Crop and livestock response may be studied for two reasons. Firstly, the aim may be the purely positivistic one of describing the response process. Except for the basic physiological phenomena explaining why response follows a particular pattern, the response process is fully described by the factor–product and factor–factor relations of Chapter 1.

The other possible purpose of response analysis is the normative one of problem solving. Rather than knowledge simply for the sake of knowledge, the normative approach aims at manipulation of the response process so as to best achieve some result. The particular normative purpose will depend on the interest of the researcher. It may range from purely physiological considerations to some economic interest. Moreover, while response research may initially be pursued by one researcher for purely positivistic purposes, the data so obtained will frequently be adapted for normative purposes by some other researcher. Normative analysis, however, with its emphasis on manipulation of the response process, requires an additional superstructure of analytical principles over and above those of Chapter 1. These additional principles vary somewhat according as to whether or not time and/or risk influences are relevant. In this chapter we will confine ourselves to the timeless and riskless analysis of response efficiency, leaving time-based considerations till Chapter 3 and risk considerations until Chapter 4.

2.2 Best Operating Conditions

Given some particular normative aim, manipulation of the response process is carried out by controlling input levels. Accordingly, BEST

OPERATING CONDITIONS are specified by the set of input quantities which best achieves the specified goal, whatever this goal may be. Given knowledge of the response function and specification of the goal, the analytical problem is to determine best operating conditions.

2.3 Output Gains and Input Losses

Any crop or livestock response process involves gains and losses. The gains are the output produced; losses are the inputs consumed. With gains and losses measured relative to some specified normative goal, best operating conditions occur when net gains are maximized. For this purpose, it is necessary for output gains and input losses to be measured in comparable units. To convert the X's and Y to comparable units, each must be weighted by an appropriate conversion factor. These conversion factors are not determined within the response process. They must be chosen *a priori* by the researcher on some basis pertinent to his purpose for needing best operating conditions.

Denoting the net gain from the response process by π and the conversion constants by p_y for Y and p_i for the variable input X_i, we have†

$$\pi = p_y Y - \sum p_i X_i \qquad (p_y, p_i \geq 0). \qquad (2.1)$$

If the weights p_y and p_i were not used to convert output and variable inputs to a common measure, it would be impossible to assess the response gains $(p_y Y)$ and losses $(\sum p_i X_i)$. Without such weights, we could not compare the net advantage of operating with alternative input levels. Conversely, some system of p's is implicit in any analysis of response which discusses the efficiency of response.

Equation (2.1) is known as the OBJECTIVE FUNCTION since it specifies π, the variable to be maximized. We can also write

$$\pi = p_y f(X_1, X_2, \ldots, X_n) - \sum p_i X_i \qquad (2.2)$$

or more simply

$$\pi = g(X_1, X_2, \ldots, X_n) \qquad (2.3)$$

† No allowance is made for losses corresponding to the consumption of fixed inputs since, being fixed, under the timeless and riskless analysis of this chapter they play no part in determining best operating conditions.

since the p's are constants. However, unlike for the response function $Y = f(X_1, X_2, \ldots, X_n)$, we always know that the objective function involves the basic algebraic form given in equation (2.1).

Two important provisos must be made about objective function (2.1). Firstly, like Y and the X's, the product weight p_y cannot be negative. If p_y were zero or less, output would always be disadvantageous. Too, while negative input weights (i.e. $p_i < 0$) may very occasionally be relevant, we will consider only the normal case of non-negative p_i values since, if need be, the logic so developed can be extended to allow for negative p_i values. The second proviso on the objective function is that frequently the goal of maximizing π will be subject to one or more CONSTRAINTS. For example, $\sum p_i X_i$ may be fixed at some level which cannot be exceeded; or only a limited amount of some variable input(s) may be available for allocation among a number of response processes. Whatever the constraints, the procedure is exactly the same in principle except that we evaluate π as a constrained maximum.

The two most frequent normative goals in crop and livestock response analysis are to determine best operating conditions for: (a) obtaining maximum physical output; and (b) obtaining maximum financial profit. Let us examine both these goals in terms of objective function (2.1), for the moment ignoring the possibility of any constraints.

Maximization of physical output implies that the objective function has the form

$$\pi = Y. \qquad (2.4)$$

This implies p values of one for p_y and zero for p_i, so that the response process involves only gains and no losses. Since in this case the objective and response functions are identical, the maximization of physical output constitutes the simplest case of determining best operating conditions.

Maximization of financial gain implies maximizing the difference between the financial value of output and the cost of the variable inputs used. If the price of output is r_y per unit and of X_i is r_i per unit, we have

$$\text{Profit} = r_y Y - \sum r_i X_i \qquad (2.5)$$

which is exactly analogous to objective function (2.1). In other words, and not surprisingly, the conversion factors for financial analysis are the market prices of the output and inputs.

So long as the research aim—whatever it be—is normative, best operating conditions can always be defined in terms of an objective function akin to equation (2.1). Whether the aim be maximum output, maximum financial gain, or something of a more fundamental nature concerned with the physiology of crop or livestock response, the only difference in analytical procedure lies in the choice of the conversion factors to be used in comparing gains and losses. These p's—whether they be physical constants, market prices, or something else decided by the researcher— always represent the measures of relative scarcity used to evaluate output gains and input losses. In this sense, regardless of whether or not the p's are money prices, the determination of best operating conditions is an economizing problem. For this reason we will refer to the p's as PRICES and π as PROFIT. Note that π includes no measure of the losses involved in consuming fixed inputs.† For the moment, this omission is immaterial since the fixed inputs do not affect the choice of best operating conditions under the timeless conditions being discussed here. However, when we introduce time and risk considerations in Chapters 3 and 4 respectively, we will have to allow for the cost of fixed inputs.

2.4 Single Variable Input

We have
$$Y = f(X_1). \tag{2.6}$$
For the unconstrained objective function we have
$$\pi = p_y Y - p_1 X_1 \tag{2.7}$$
and, taking the first derivative of objective function (2.7),
$$\partial \pi / \partial X_1 = p_y (\partial Y / \partial X_1) - p_1. \tag{2.8}$$

To maximize profit, we set $\partial \pi / \partial X_1$ equal to zero and solve for X_1. The necessary second-order condition $(\partial^2 \pi / \partial X_1^2 < 0)$ holds automatically from the assumption of diminishing returns. Setting equation (2.8) equal to zero, the profit maximizing condition is
$$\partial Y / \partial X_1 = p_1 / p_y. \tag{2.9}$$

† In this chapter π thus corresponds to the gross margin per technical unit (e.g. per hectare or per animal).

Thus with a single variable input and no constraints on the objective function, best operating conditions occur when the marginal product of the input equals the INVERSE PRICE RATIO p_1/p_y.

Since p_1 and p_y must be non-negative, equation (2.9) implies $\partial Y/\partial X_1$ can never be negative at the level of X_1 that maximizes profit. Regardless of the size of p_1 and p_y, any level of X_1 where MP_1 is negative will always be irrational since a higher level of output could be achieved with fewer inputs.

The profit maximizing condition of equality between $\partial Y/\partial X_1$ and p_1/p_y is shown in Fig. 2.1. The line AB with slope p_1/p_y is constructed tangential to the response curve OB so that equation (2.9) is satisfied at point B. Optimal input is OC, yielding output of OD. Profit is p_y times OA since

$$\pi = p_y(\text{OD}) - p_1(\text{OC})$$
$$= p_y(\text{OD}) - p_y(\text{AD})$$
$$= p_y(\text{OA}).$$

Operation at any point other than B would yield a smaller profit under the given price régime. For example, operation at E yields profit of $p_y(\text{OF})$ which is clearly less than $p_y(\text{OA})$.

In Fig. 2.1, for the given values of p_1 and p_y, lines AB and FE are but

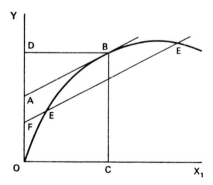

Fig. 2.1. Iso-profit lines and best operating conditions for a single-variable input response process.

two members of the family of ISO-PROFIT LINES obtained by rearranging the objective function in the form:

$$Y = \pi/p_y + (p_1/p_y)X_1. \qquad (2.10)$$

For given values of π, p_1, and p_y, equation (2.10) is the *locus* of all combinations of X_1 and Y that yield the given π. Hence AB of Fig. 2.1 can be interpreted as the highest iso-profit line achievable under the given conditions of $Y = f(X_1)$, p_1 and p_y.

An intuitive justification of the profit maximizing condition of equation (2.9) is obtained by expressing the equation in incremental units and rearranging it as

$$p_y \Delta Y = p_1 \Delta X_1. \qquad (2.11)$$

This equality implies that the cost of the last increment of input used should be just equal to the value of the extra output obtained by using that increment of input. In other words, for maximum profit, the last unit of input must just pay for itself. Because of diminishing returns, prior units of input will have more than paid for themselves; further units of input will not cover their cost.

Example for $Y = f(X_1)$

Suppose the response curve is

$$Y = 1 + 10X_1 - 2X_1^2.$$

Corresponding to equation (2.9), best operating conditions imply

$$p_1/p_y = 10 - 4X_1.$$

Solving this equation for a few different price ratios, optimal levels of X_1 and Y are as follows:

p_1/p_y	X_1	Y
10	0·0	1·0
8	0·5	5·5
6	1·0	9·0
4	1·5	11·5
2	2·0	13·0
0	2·5	13·5

36   THE ANALYSIS OF RESPONSE IN CROP AND LIVESTOCK PRODUCTION

This data exemplifies the general rule that as p_1 decreases relative to p_y, best operating conditions imply the use of more X_1. In the limit when p_1/p_y is at its minimum of zero, X_1 must be used at that level which drives its marginal product to zero. This input level, of course, corresponds to Y being at its maximum.

For a particular price ratio, say p_1/p_y equal to 6·0, the following data shows how profit changes as X_1 varies:

p_1/p_y	X_1	Y	π
6	0·0	1·0	p_y
6	0·5	5·5	$2·5p_y$
6	1·0	9·0	$3·0p_y$
6	1·5	11·5	$2·5p_y$
6	2·0	13·0	$1·0p_y$
6	2·5	13·5	$-1·5p_y$

Note that under the given price ratio, maximum profit occurs at an input level well below that which produces maximum output.

2.5 Two Variable Inputs

We have

$$Y = f(X_1, X_2) \qquad (2.12)$$

and the unconstrained objective function is

$$\pi = p_y Y - (p_1 X_1 + p_2 X_2). \qquad (2.13)$$

Maximization of π with respect to the two variable inputs implies simultaneous solution of the two equations

$$\partial \pi / \partial X_1 = 0, \qquad (2.14a)$$

$$\partial \pi / \partial X_2 = 0, \qquad (2.14b)$$

to find the combination of X_1 and X_2 that specifies best operating conditions. The required second-order condition for a maximum (that the differential $d^2\pi$ be negative) is automatically satisfied through the assumptions of diminishing and decreasing returns for the response

EFFICIENCY IN RESPONSE 37

function. Taking the first derivatives of equation (2.13), equations (2.14) can be rewritten as:

$$\partial Y/\partial X_1 = p_1/p_y, \qquad (2.15a)$$

$$\partial Y/\partial X_2 = p_2/p_y. \qquad (2.15b)$$

Analogous to the case of a single variable input, the profit maximizing condition specified by equations (2.15) is that the marginal product of each factor (MP_i) must equal the relevant inverse price ratio (p_i/p_y).

Irrational Input Levels and the Relevant Range

For $Y = f(X_1)$, it was shown that since p_i/p_y must be non-negative, X_1 levels for which MP_1 is negative are always irrational; if MP_1 is negative, reducing X_1 until $\partial Y/\partial X_1$ is zero must always increase profit. Likewise, since p_1, p_2 and p_y must be non-negative, equations (2.15) indicate maximum profit with two variable inputs can never occur where either MP_1 or MP_2 is negative. Relative to Fig. 1.8, best operating conditions can only occur within the region OARB of positive marginal products or diminishing negative rates of technical substitution. The obvious logic of this is that for any X_1, X_2 combination with MP_1 or MP_2 negative, the same output can always be obtained with less of one or both inputs. For example, in Fig. 2.2 any X_1, X_2 combination outside

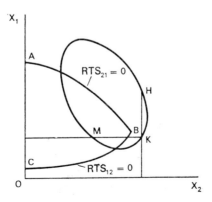

Fig. 2.2. Regions of rational and irrational input combinations.

the region ABC bounded by the ridge-lines AB and BC would be irrational. For any points H or K outside this region of diminishing negative RTS_{12} another point M, on the same isoquant but in the region of positive marginal products can be found where less of one or both inputs would suffice to produce the same output. Accordingly, the input ranges spanned by the region of non-negative marginal products constitute the relevant range for normative response analysis. Best operating conditions can never lie outside these input ranges.

Example for $Y = f(X_1, X_2)$

As in Section 1.6, suppose the response function is

$$Y = 18 \cdot 846512 + 7 \cdot 586167N + 2 \cdot 469969P - 0 \cdot 655713N^2$$
$$- 0 \cdot 397513P^2 + 0 \cdot 211423NP. \qquad (2.16)$$

Setting MP_N and MP_P equal to the relevant inverse price ratios, we have respectively:

$$p_N/p_y = 7 \cdot 586167 - 1 \cdot 311426N + 0 \cdot 211423P, \qquad (2.17a)$$

$$p_P/p_y = 2 \cdot 469969 - 0 \cdot 795026P + 0 \cdot 211423N. \qquad (2.17b)$$

Simultaneous solution of these two equations yields the optimal combinations of N and P, as shown for a few different price ratios in Table 2.1 along with the corresponding levels of output and profit.

As dictated by common sense, the data of Table 2.1 illustrate that best operating conditions imply:

(i) increased input and output as inputs become cheaper relative to output;

(ii) increased use of an input as its price decreases relative to the price of other inputs;

(iii) increased profit as input prices decline relative to the price of output; and

(iv) dependence of best operating conditions on the ratios of input and output prices, not on the absolute level of the individual prices. For example, doubling of all prices leads to no change in best operating conditions.

TABLE 2.1
VARIATIONS IN BEST OPERATING CONDITIONS FOR RESPONSE FUNCTION (2.16)

p_P/p_N	p_N/p_y	p_P/p_y	N	P	Y	π
2·00	0·5	1·0	5·96	3·43	48·70	42·29p_y
3·00	0·5	1·5	5·88	2·76	48·00	40·92p_y
4·00	0·5	2·0	5·77	2·10	46·79	39·70p_y
1·00	1·0	1·0	5·58	3·31	48·48	39·59p_y
1·50	1·0	1·5	5·45	2·67	47·56	38·10p_y
2·00	1·0	2·0	5·37	2·00	46·30	36·93p_y
0·67	1·5	1·0	5·57	3·20	48·36	36·80p_y
1·00	1·5	1·5	5·08	2·55	46·91	35·46p_y
1·33	1·5	2·0	4·95	1·90	45·57	34·34p_y

2.6 n Variable Inputs

With n variable inputs, choice of best operating conditions is simply an extension of the procedure for two variable inputs. We have

$$Y = f(X_1, X_2, \ldots, X_n) \qquad (2.18)$$

and the unconstrained objective function is

$$\pi = p_y Y - \sum p_i X_i. \qquad (2.19)$$

Maximization of π with respect to the n input variables implies simultaneous solution of the n equations

$$\partial \pi / \partial X_i = 0 \qquad (i = 1, 2, \ldots, n). \qquad (2.20)$$

Given the algebraic form of the objective function and the assumptions of diminishing and decreasing returns, the second-order condition for a maximum (that the differential $d^2\pi$ be negative) is automatically satisfied. Each of the n equations (2.20) can be rearranged as

$$MP_i = p_i/p_y. \qquad (2.21)$$

Solution of this system of n equations yields the set of X_i's that constitute best operating conditions. Substitution of these values into the response and objective functions, respectively, gives the level of output and profit under best operating conditions. Note that since p_i and p_y must be non-negative, equations (2.21) indicate MP_i must always be non-negative for profit maximization. As with one or two variable inputs, production with any MP_i negative is always irrational. Correspondingly, the relevant range for response analysis of n variable inputs consists of the input ranges spanning the surface area where all marginal products are non-negative.

Example for $Y = f(X_1, X_2, \ldots, X_n)$

For the second degree polynomial response function with n variable inputs,

$$Y = a_0 + \sum a_i X_i + \sum a_{ii} X_i^2 + \sum\sum a_{ij} X_i X_j \quad (i < j) \qquad (2.22)$$

profit maximization implies simultaneous solution of the n equations of the form

$$p_i/p_y = a_i + 2a_{ii}X_i + \sum_j a_{ij} X_j \qquad (2.23)$$

corresponding to equation set (2.21).

For the power function

$$Y = \Pi X_i^{b_i} \qquad (\sum b_i < 1, b_i > 0) \qquad (2.24)$$

we have

$$\pi = p_y \Pi X_i^{b_i} - \sum p_i X_i \qquad (2.25)$$

and best operating conditions are specified by the set of n equations

$$p_i/p_y = b_i Y/X_i. \qquad (2.26)$$

2.7 Multiple Response without Input Control

Frequently with crops and livestock, the response process may simultaneously produce a variety of outputs. For example, we have grain and straw in cereals; bone, muscle and fat in livestock; or various grades of meat in beef production. In such cases it is impossible to allocate inputs between the various responses, that being done within the plant or animal.

Suppose there are r simultaneous responses. Each may be characterized by a "response function"

$$Y_k = f_k(X_1, X_2, \ldots, X_n) \qquad (k = 1, 2, \ldots, r) \qquad (2.27)$$

where X_i is the total amount of the ith input available to the plant or animal for internal allocation between the various Y's.

The objective function is

$$\pi = \sum p_k Y_k - \sum p_i X_i \qquad (2.28)$$

and subject to the second-order condition that the differential $d^2\pi$ be negative, best operating conditions imply simultaneous solution of the set of n equations for $\partial \pi / \partial X_i$ equal to zero. These n equations are:

$$\sum_k p_k(\partial Y_k / \partial X_i) - p_i = 0. \qquad (2.29)$$

Example

For a 1963 potato-fertilizer trial at San Pablo in the Central Valley, Chile, the response functions for three grades of potatoes G_1, G_2 and G_3 making up the total output were estimated to be:

$$G_1 = 13 \cdot 519340 + 2 \cdot 157720N + 1 \cdot 620792P - 0 \cdot 107980N^2$$
$$- 0 \cdot 212495P^2 + 0 \cdot 085939NP, \qquad (2.30)$$

$$G_2 = 7 \cdot 610232 - 0 \cdot 084552N - 0 \cdot 347490P + 0 \cdot 000251N^2$$
$$+ 0 \cdot 063609P^2 + 0 \cdot 022813NP, \qquad (2.31)$$

$$G_3 = 3 \cdot 088811 - 0 \cdot 036608N - 0 \cdot 055834P - 0 \cdot 000084N^2$$
$$+ 0 \cdot 014167P^2 + 0 \cdot 010313NP. \qquad (2.32)$$

In these equations, G_1, G_2 and G_3 are, respectively, yields of large, medium and small tubers in tonnes per hectare; N is potassium nitrate in units of 150 kilograms per hectare; and P is triple superphosphate in units of 100 kilograms per hectare.

Corresponding to equation (2.29), for p_N of 1·0, p_P of 2·0, and G_1, G_2, G_3 with respective prices of 5·0, 4·0 and 3·0, best operating conditions occur for the N and P values that satisfy the pair of equations:

$$9·340568 - 1·078296N + 0·551886P = 0, \quad (2.33a)$$

$$4·546498 + 0·551886N - 1·531076P = 0. \quad (2.33b)$$

Solution of these equations gives the optimal input array of 12·49 units of N and 7·47 units of P.

2.8 Multiple Response with Input Control

Consider now the case of an array of independent response processes for each of which control can be exercised over the level of inputs.

Suppose there are r response processes specified by the r response functions

$$Y_h = f_h(X_{1h}, X_{2h}, \ldots, X_{nh}) \quad (h = 1, 2, \ldots, r) \quad (2.34)$$

where X_{ih} is the quantity of X_i allocated to the hth response process and may be always zero for some Y_h's. We can write the overall unconstrained objective function as

$$\pi = \sum \pi_h \quad (2.35)$$

where π_h is the profit from the hth response process. Because the response processes are independent (Y_h depends only on X's, not on any other Y's), we must have

$$\max \pi = \sum (\max \pi_h). \quad (2.36)$$

Accordingly, best operating conditions for the overall array of response processes occur when each individual process considered in isolation is at best operating conditions. Hence, from Section 2.6, overall best operating conditions occur when the r independent sets of n (or less if some X's are always zero in some processes) equations

$$\partial Y_h/\partial X_i = p_i/p_h \quad (2.37)$$

hold true. Alternatively stated, by independently solving each of these r sets of n simultaneous equations, we can derive the r sets of n input levels

$(X_{1h}, X_{2h}, \ldots, X_{nh})$ which specify best operating conditions. The required second-order condition for a maximum (that the differential $d^2\pi$ be negative) is again automatically satisfied through the response assumptions of diminishing and decreasing returns.

Since p_i/p_h must be non-negative, equation (2.37) indicates that any input array for which one or more marginal products are negative must be irrational. Furthermore, as shown relative to Fig. 1.8, it follows that all RTS_{ij}'s must be negative and decreasing at best operating conditions. Rearranging equation (2.37), we have the set of rn equalities

$$1 = p_h(MP_{ih})/p_i \quad (h = 1, 2, \ldots, r; i = 1, 2, \ldots, n) \quad (2.38)$$

indicating that for maximization of π, the value of the marginal product of X_i in the hth process $(p_h MP_{ih})$ must equal p_i. By manipulating the equalities (2.38), we can also see that best operating conditions for the set of r response processes imply the following relationships:

(i) *Factor–product.* In each response process, the marginal product of X_i, MP_{ih}, must equal the factor–product price ratio p_i/p_h, as specified by equation (2.37).

(ii) *Factor–factor.* In each response process, $RTS_{ij} (= -MP_{jh}/MP_{ih})$ must equal the negative of the inverse factor–factor price ratio p_j/p_i since equation (2.38) implies

$$p_h(MP_{ih})/p_i = p_h(MP_{jh})/p_j.$$

(iii) *Product–product.* For any two products Y_h and Y_k, the RATE OF TECHNICAL TRANSFORMATION or substitution of Y_h for Y_k, written RTT_{hk} and given by $\partial Y_h/\partial Y_k$ or $-MP_{ih}/MP_{ik}$, must equal the negative of the inverse product–product price ratio p_k/p_h since equation (2.38) implies

$$p_h(MP_{ih})/p_i = p_k(MP_{ik})/p_i.\dagger$$

† Because of the assumption of diminishing and decreasing returns, RTT_{hk} will always be negative and increasing (like RTS_{ij} on segment CD of Fig. 1.6). If our response theory of Sect. 1.2 allowed for an initial stage of increasing returns, it would be necessary to specify that RTT_{hk} be negative and increasing (rather than decreasing) for best operating conditions. The fact that RTT_{hk} must be negative within the relevant range is intuitively obvious since as we substitute Y_h for Y_k at the margin, we gain MP_{ih} and lose MP_{ik} so that the ratio MP_{ih}/MP_{ik} must have a negative sign placed on it. See Dorfman (1964, ch. 6).

2.9 Constraints on the Objective Function

So far we have assumed no constraints on the objective function. Constraints can be important. They may take two main forms. Firstly, the level of output may be fixed so that best operating conditions devolve to finding the input array that has the LEAST COST (i.e. $\sum p_i X_i$ a minimum) for the required level of output. Secondly, the total outlay $\sum p_i X_i$ may be limited so that best operating conditions devolve to finding the most profitable input array under the outlay limitation.

Variations on these two main types of constraint may occur. For instance, only some output levels may be fixed in a series of response processes; or outlay may only be limited for some subset of the variable inputs. But whatever the constraint variation, the same principles as outlined below for the general least-cost and fixed outlay constraints are applicable.

2.9.1. FIXED-OUTPUT CONSTRAINTS

In sequence, we will consider three types of output constraint:

(i) single response process, output (Y) fixed;
(ii) r response processes, each with output (Y_h) fixed;
(iii) r response processes, total returns ($\sum p_h Y_h$) fixed.

Single Response Process, Y Fixed

Suppose the response function is

$$Y = f(X_1, X_2, \ldots, X_n) \tag{2.39}$$

and that output must be at some level Y^*. This constraint is entered in the objective function via the Lagrangian multiplier λ so that we have the constrained objective function

$$\pi = p_y Y - \sum p_i X_i + \lambda(Y - Y^*). \tag{2.40}$$

Setting $\partial \pi / \partial X_i$ and $\partial \pi / \partial \lambda$ equal to zero to satisfy the first-order conditions for the constrained maximum, we have the $n + 1$ equations:

$$p_y(\partial Y/\partial X_i) - p_i + \lambda(\partial Y/\partial X_i) = 0, \tag{2.41a}$$

$$Y - Y^* = 0. \tag{2.41b}$$

EFFICIENCY IN RESPONSE 45

From equation (2.41a) we have

$$\lambda = p_i(\partial X_i/\partial Y) - p_y. \qquad (2.42)$$

Eliminating λ and p_y from these n equations gives the $n - 1$ relationships:

$$p_i/p_j = MP_i/MP_j \qquad (i \neq j). \qquad (2.43)$$

Since p_i/p_j must be non-negative, MP_i and MP_j must both be either positive or negative. Hence equations (2.43) correspond to the $n - 1$ (isocline) equations

$$RTS_{ij} = -p_j/p_i \qquad (2.44)$$

lying in the surface regions of negative (decreasing or increasing) RTS_{ij}. These $n - 1$ equations (2.44), together with equation (2.41b) rearranged as the isoquant equation,

$$X_1 = f(X_2, X_3, \ldots, X_n; Y^*), \qquad (2.45)$$

constitute n equations whose simultaneous solution yields the n input levels for least-cost production of Y^*—subject to two provisos.

The first proviso relates to second-order conditions for π to be maximized. If the response function allows negative marginal products, then the n equations will yield two sets of input arrays. There will be a least-cost array in the region of positive marginal products, and a maximum-cost array in the region of negative marginal products. Both input arrays will lie on the Y^* isoquant. In such cases the least-cost solution is that which satisfies the second-order condition that all RTS_{ij} be negative and decreasing (or, same thing, that all MP_i's be positive). The easiest check is to calculate the cost of each input array; the one with the smaller cost is the desired set.

The second proviso on the least-cost array is that no X_i can be negative. It may be that the arithmetic solution of equations (2.44) and (2.45) gives one or more negative input levels. The feasible least-cost array must then include some boundary solutions, which simply means that one or more inputs must be used at zero level. To find this feasible least-cost array if all X_i are negative, it is necessary to compare the cost of each of the input arrays involving the intersection of the Y^* isoquant with the input axes. If only some X_i are negative, it is the intersection of the Y^*

isoquant with the $X_i = 0$ axes that is relevant. What this amounts to is that between the origin and the intersection of an isocline with an input axis, the input axis itself serves as the isocline. For example, in Fig. 2.3 the isocline for $\partial N/\partial P$ equal to $-1\cdot 5$ is the path OAG.

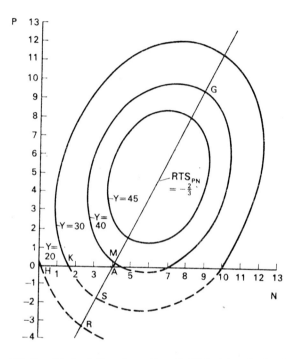

Fig. 2.3. Graphical solution of equations (2.46) to determine least-cost input levels.

Since the isocline equations (2.44) and the isoquant equation (2.45) must be satisfied simultaneously, it follows that least-cost production of Y^* is specified by the intersection of the isoclines with the isoquant. Accordingly, as Y^* increases, the isoclines trace out the path of least-cost input combinations under the given price régime. For this reason isoclines are sometimes called LEAST-COST EXPANSION PATHS.

Example

For the El Llano wheat-fertilizer response function (2.16), what combination of N and P would be least cost, and hence most efficient for a yield of Y^* quintals if p_N/p_y were 1·0 and p_P/p_y were 1·5?

Corresponding to equations (2.44) and (2.45), the least-cost input combination is given by solution of the n (equals two) equations consisting of the isocline or least-cost expansion path

$$-1\cdot 5 = -(2\cdot 469969 - 0\cdot 795026P + 0\cdot 211423N)/$$
$$(7\cdot 586167 - 1\cdot 311426N + 0\cdot 211423P) \qquad (2.46a)$$

and the isoquant function

$$Y^* = 18\cdot 846512 + 7\cdot 586167N + 2\cdot 469969P$$
$$- 0\cdot 655713N^2 - 0\cdot 397513P^2 + 0\cdot 211423NP. \qquad (2.46b)$$

Because of the quadratic nature of response function (2.16) (thereby allowing both positive and negative marginal products), equation (2.46b) has two roots. In consequence there are two solutions to equations (2.46). That for which $p_N N$ plus $p_P P$ is the smaller is the required least-cost combination, the other solution being the maximum-cost input combination for Y^*. Solving equations (2.46) for a few Y^* values gives the following least-cost input combinations:

Y^*:	20	30	40
N:	2.464	3.224	4.218
P:	−3.185	−1.696	0.252

Since negative input levels are infeasible, for output levels of 20 and 30 it is necessary to consider the boundary solutions giving the required yields with P equal to zero. In other words, for Y^* of 20 and 30, the relevant isocline segment is not equation (2.46a) but the input axis

$$P = 0. \qquad (2.46c)$$

Solving equations (2.46b) and (2.46c) for Y^* of 20 and 30 to obtain the feasible input solutions, we have the following required least-cost input levels:

Y^*:	20	30	40
N:	0·155	1·728	4·218
P:	0	0	0·252

These feasible least-cost combinations correspond respectively to the points H, K and M where the expansion path OAG of Fig. 2.3 intersects the 20, 30 and 40 quintal isoquants. The infeasible least-cost solutions are shown at points R and S. Indeed, prior inspection of Fig. 2.3 would have indicated that P must be zero for least-cost output of 20 or 30. The maximum-cost input combination for Y^* of 40 occurs at point G.

r Response Processes, Each with Output Y_h Fixed

Consider now a series of r independent response processes

$$Y_h = f_h(X_{1h}, X_{2h}, \ldots, X_{nh}) \quad (h = 1, 2, \ldots, r) \quad (2.47)$$

to be operated under the constraints:

$$Y_h = Y^*_h. \quad (2.48)$$

The constrained objective function is

$$\pi = \sum p_h Y_h - \sum\sum p_i X_{ih} + \sum \lambda_h (Y_h - Y^*_h) \quad (2.49)$$

where X_{ih} is the quantity of X_i used in the hth response process. Applying the usual procedure to maximize π with respect to X_{ih} and λ gives r independent sets of n equations. Each of these sets corresponds to equations (2.44) and (2.45). With appropriate accommodation of any required boundary solutions, simultaneous solution of each of these r sets (subject to the second-order condition that all MP_i's be non-negative) gives the required r sets of least-cost input arrays.

r Response Processes, Total Return $\sum p_h Y_h$ Fixed

We now desire best operating conditions for response functions (2.47) subject to the constrained objective function:

$$\pi = \sum p_h Y_h - \sum\sum p_i X_{ih} + \lambda(\sum p_h Y_h - R) \quad (2.50)$$

where R is the prespecified fixed level of total revenue $\sum p_h Y_h$.

EFFICIENCY IN RESPONSE 49

Setting $\partial \pi / \partial X_{ih}$ and $\partial \pi / \partial \lambda$ equal to zero, we have the $m + 1$ first-order conditions:

$$p_h(\partial Y_h/\partial X_i) - p_i + \lambda p_h(\partial Y_h/\partial X_i) = 0, \quad (2.51a)$$

$$R - \sum p_h Y_h = 0. \quad (2.51b)$$

Eliminating λ from the m equations (2.51a) gives the $m - 1$ equations

$$(p_{y1}/p_1)(\partial Y_1/\partial X_1) = (p_h/p_i)(\partial Y_h/\partial X_i), \quad (2.52)$$

where $h = 1, 2, \ldots, r$; $i = 1, 2, \ldots, n$; and $i \neq 1$ if $h = 1$. These $m - 1$ equations together with equation (2.51b) constitute the set of m equations to be solved simultaneously for the m input levels constituting least-cost operating conditions for a total return of R. Should the solution involve any negative X_{ih}'s, boundary solutions involving some X_{ih} equal to zero must be explored to find the feasible least-cost array.

The necessary second-order condition for the input array to be least-cost and hence maximize π (rather than a maximum-cost array minimizing π) is that MP_i be non-negative at each of the m input levels. This necessity can be seen from equation (2.52). Since prices must be non-negative, these equations only specify that all MP_i have the same sign; they could be negative which would mean irrational (cost maximizing!) input levels.

The m equations (2.51b) and (2.52) can be categorized into four types of (first-order) conditions required for least-cost production of a number of outputs to give a fixed total revenue R. These conditions are as follows:

(i) The ISO-REVENUE LOCUS

$$Y_1 = (R/p_{y1}) - \sum (p_k/p_{y1}) Y_k \quad (k = 2, 3, \ldots, r), \quad (2.53)$$

obtained by rearranging equation (2.51b), must be satisfied. This (linear) equation locates all output combinations with a total value of R. (An example for the case of two outputs is the line BC of Fig. 2.5 below.)

(ii) The $r - 1$ product–product relations

$$RTT_{1k} = -p_k/p_{y1} \quad (k = 2, 3, \ldots, r) \quad (2.54)$$

must be satisfied. These relations imply that the rate of technical transformation between any two products must equal their negative inverse price ratio. Equations (2.54) are analogous to isoclines except that they lie in output-output space. Subject to satisfaction of the other required conditions—(iii) and (iv) below, and all MP_i's non-negative—equations (2.54) specify the least-cost expansion path in output space. (An example for the case of two outputs is the line OA of Fig. 2.5 below.)

(iii) The $n - 1$ factor–factor relations

$$RTS_{1j} = -p_j/p_1 \quad (j = 2, 3, \ldots, n) \tag{2.55}$$

must hold true. These isocline equations specify that the rate of technical substitution between any two factors must equal their negative inverse price ratio. Hence they trace out the least-cost expansion paths in input space.

(iv) The $(n - 1)(r - 1)$ factor–product relations

$$p_{y1}(MP_{11})/p_1 = p_k(MP_{jk})/p_j \quad (j = 2, 3, \ldots, n;$$
$$k = 2, 3, \ldots, r) \tag{2.56}$$

must be satisfied, where MP_{jk} is the marginal product of X_j in producing Y_k. These relations imply that the ratio of (a) the value of the marginal product of any factor in any response process to (b) the price of the factor, must be the same for all factors and products. This requirement is analogous to the unconstrained case for r processes except that the ratios of equation (2.56), unlike the ratios implied by equation (2.38), need not be equal to unity.

Example

Given the following response functions for wheat on the two different soil types of El Llano and Pirque in the Central Valley, Chile, how should inputs of N and P be distributed between the two soil types to best achieve a total revenue of 1000 per hectare if p_N is 6, p_P is 8 and p_y is 10? The response functions are:

$$Y_1 = 18 \cdot 846512 + 7 \cdot 586167N + 2 \cdot 469969P - 0 \cdot 655713N^2$$
$$- 0 \cdot 397513P^2 + 0 \cdot 211423NP, \tag{2.57}$$

$$Y_2 = 29 \cdot 988229 + 1 \cdot 491994N + 1 \cdot 107624P - 0 \cdot 180426N^2$$
$$- 0 \cdot 004586P^2 + 0 \cdot 090326NP, \tag{2.58}$$

where Y_1 and Y_2 are quintals of wheat per hectare; N is sodium nitrate in units of 150 kilograms per hectare; and P is triple superphosphate in units of 100 kilograms per hectare.

The constrained objective function corresponding to equation (2.50) is

$$\pi = 10(Y_1 + Y_2) - 6(N_1 + N_2) - 8(P_1 + P_2)$$
$$+ \lambda(10Y_1 + 10Y_2 - 1000). \qquad (2.59)$$

Differentiating π with respect to N_h, P_h and λ, we obtain the following set of five (i.e. $nr + 1$) equations to be set equal to zero and solved for best-operating conditions.

$$\partial \pi / \partial N_1 = 10(\partial Y_1 / \partial N_1) - 6 + 10\lambda(\partial Y_1 / \partial N_1), \qquad (2.60a)$$
$$\partial \pi / \partial N_2 = 10(\partial Y_2 / \partial N_2) - 6 + 10\lambda(\partial Y_2 / \partial N_2), \qquad (2.60b)$$
$$\partial \pi / \partial P_1 = 10(\partial Y_1 / \partial P_1) - 8 + 10\lambda(\partial Y_1 / \partial P_1), \qquad (2.60c)$$
$$\partial \pi / \partial P_2 = 10(\partial Y_2 / \partial P_2) - 8 + 10\lambda(\partial Y_2 / \partial P_2), \qquad (2.60d)$$
$$\partial \pi / \partial \lambda = 10Y_1 + 10Y_2 - 1000. \qquad (2.60e)$$

Setting these equations equal to zero and eliminating λ from the first four, we obtain the following four equations corresponding to equations (2.52) and (2.53):

$$(10/6)(\partial Y_1 / \partial N_1) = (10/6)(\partial Y_2 / \partial N_2), \qquad (2.61a)$$
$$(10/6)(\partial Y_1 / \partial N_1) = (10/8)(\partial Y_1 / \partial P_1), \qquad (2.61b)$$
$$(10/6)(\partial Y_1 / \partial N_1) = (10/8)(\partial Y_2 / \partial P_2), \qquad (2.61c)$$
$$Y_1 = 100 - Y_2. \qquad (2.61d)$$

Equation (2.61d) is the iso-revenue equation (2.53); equations (2.61 a, b, c) in turn correspond to the product–product, factor–factor, and factor–product requirements of equations (2.54), (2.55) and (2.56). Making the appropriate substitutions for Y_h and MP_{ih} from the parent response functions and rearranging, equations (2.61) become:

$$P_1 = 6 \cdot 202854 N_1 - 1 \cdot 706769 N_2 + 0 \cdot 427227 P_2$$
$$- 28 \cdot 824410, \qquad (2.62a)$$
$$P_1 = 1 \cdot 819987 N_1 - 7 \cdot 096525, \qquad (2.62b)$$

$$P_1 = 6{\cdot}202854N_1 - 0{\cdot}320430N_2 - 0{\cdot}032537P_2$$
$$- 31{\cdot}929919, \tag{2.62c}$$
$$O = 0{\cdot}397513P_1{}^2 - (2{\cdot}469969 + 0{\cdot}211423N_1)P_1$$
$$+ (51{\cdot}165259 - 7{\cdot}586167N_1 - 1{\cdot}491994N_2$$
$$- 1{\cdot}107624P_2 + 0{\cdot}655713N_1{}^2 + 0{\cdot}180426N_2{}^2$$
$$+ 0{\cdot}004586P_2{}^2 - 0{\cdot}090326N_2P_2). \tag{2.62d}$$

Solving these four equations, we obtain the required least-cost input allocation for a total revenue of 1000. This solution is that 5·27 and 2·49 units of N and P respectively should be allocated to Y_1 (giving a yield of 47·11), while 4·62 and 13·57 units of N and P respectively should be allocated to Y_2 (giving a yield of 52·89). As required for the solution, $10(Y_1 + Y_2)$ equals 1000.

The situation is illustrated diagrammatically in Figs. 2.4 and 2.5. Drawn in the input plane, Figs. 2.4(a) and (b) show for Y_1 and Y_2, respectively, the least-cost isocline and the intersection at point A of this isocline with the output isoquant which satisfies the requirements of equations (2.61 a, c and d). In Fig. 2.4(a) the point B (with N at 7·86 and P at 7·21) corresponds to the N_1 and P_1 parts of the second (i.e. maximum cost) solution of equations (2.62). Figure 2.5, drawn in the output plane, shows the intersection of the iso-revenue line BC of equation (2.62d) with the least-cost expansion path OA specified by equations (2.62 a, b and c). Figure 2.4 completely specifies best operating conditions for a total revenue of 1000 under the given price régime. In contrast, Fig. 2.5 does not. It only shows the quantity of Y_1 and Y_2 to be produced, without indicating how inputs should be allocated within each process.

2.9.2. FIXED-OUTLAY CONSTRAINTS

In sequence, we will consider three types of input constraint:
(i) single response process, outlay $(\sum p_i X_i)$ limited;
(ii) r response processes, outlay $(\sum_i p_i X_{ih})$ limited for each individual process;
(iii) r response processes, overall outlay $(\sum\sum p_i X_{ih})$ limited but outlay not fixed for each individual process.

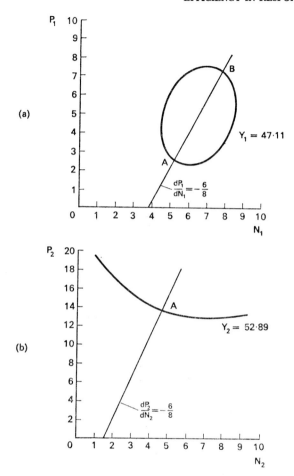

FIG. 2.4. Part-solution of equations (2.62) to determine best input allocation between (a) response process (2.57) and (b) response process (2.58).

Single Response Process, $\sum p_i X_i$ Limited

Suppose the response function is

$$Y = f(X_1, X_2, \ldots, X_n) \tag{2.63}$$

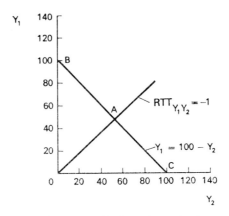

Fig. 2.5. Part-solution of equations (2.62) to determine best output levels of response process (2.57) and (2.58) in achieving a given total revenue.

and that total outlay $\sum p_i X_i$ can be no greater than some quantity C. For outlay fixed at C, we have the constrained objective function

$$\pi = p_y Y - \sum p_i X_i + \lambda(\sum p_i X_i - C). \quad (2.64)$$

Setting $\partial \pi / \partial X_i$ and $\partial \pi / \partial \lambda$ equal to zero gives the $n + 1$ equations:

$$p_y(\partial Y/\partial X_i) - p_i + \lambda p_i = 0, \quad (2.65a)$$

$$\sum p_i X_i - C = 0. \quad (2.65b)$$

Eliminating λ from equations (2.65a) gives the $n - 1$ isocline equations

$$RTS_{ij} = -p_j/p_i \quad (i \neq j). \quad (2.66a)$$

These, together with equation (2.65b) rearranged as the ISO-COST LOCUS

$$X_1 = C/p_1 - \sum(p_j/p_1)X_j \quad (j = 2, 3, \ldots, n) \quad (2.66b)$$

constitute the n equations whose simultaneous solution yields the n input levels whose combined cost is C and which give the highest possible output (and hence profit) for that outlay. Should any X_i be negative in this solution, then boundary solutions involving the intersection of the iso-cost locus with the isocline segments $X_i = 0$ must be explored.

It remains to check that greater profit cannot be obtained for some outlay less than C. If so, then the constraint is not an effective one and the unconstrained best-operating conditions are best. The easiest check is to calculate the ratio $p_y(MP_i)/p_i$ for one of the input levels derived from equations (2.66). If this ratio is less than unity, it implies the constrained maximum involves a total outlay greater than the unconstrained solution. The ordinary procedure for an unconstrained set of best operating conditions must then be applied. The relevance of the ratio $p_y(MP_i)/p_i$ can be seen by comparing equation (2.38) for the unconstrained case with the equation

$$1 - \lambda = p_y(MP_i)/p_i \qquad (2.67a)$$

derived from equation (2.65a) for the constrained case. If $1 - \lambda$ is less than unity, it means that the last unit of X_i is not covering its cost and that the unconstrained best level has been exceeded.

Notice that there is only a single difference between solving for the fixed output and fixed outlay constraints. For fixed output, maximum profit is specified by the intersection of the least-cost isocline with the appropriate isoquant segment. For fixed outlay, it is the intersection of the least-cost isocline with the relevant iso-cost locus.

The above analysis covers the situation where the total outlay limitation on the response process is known. Often this limit to $\sum p_i X_i$ will not be known specifically but be determined by the cost of credit or by profit possibilities available in alternative processes, i.e. by the opportunity cost that would occur if better profit possibilities elsewhere were foregone. Suppose we wish to decide on outlay on a particular response process knowing that the cost of outlay per unit or the highest net return per unit of outlay available from our other processes is r. With outlay limited over all activities, outlay on X_i in the particular process under study should be restricted to that level which yields a marginal profit of r. The criterion for best operating conditions is thus

$$\partial \pi / \partial (p_i X_i) = r \qquad (i = 1, 2, \ldots, n) \qquad (2.67b)$$

which reduces to

$$\partial Y/\partial X_i = p_i(1 + r)/p_y \qquad (2.67c)$$

or, equivalent to equation (2.67a) with r equal to $-\lambda$,

$$(1 + r) = p_y(MP_i)/p_i. \qquad (2.67d)$$

Example

Again for the El Llano wheat-fertilizer response function (2.16), what combination of N and P would be most profitable, and hence most efficient, if total outlay $(p_N N + p_P P)$ were limited to 30; p_N being 4, p_P being 6 and p_y being 5?

Corresponding to equation (2.64), the constrained objective function is

$$\pi = 5Y - 4N - 6P + \lambda(4N + 6P - 30) \qquad (2.68)$$

so that for the best-operating conditions of equations (2.66) we have:

$$-1 \cdot 5 = -(2 \cdot 469969 - 0 \cdot 795026P + 0 \cdot 211423N)/$$
$$(7 \cdot 586167 - 1 \cdot 311426N + 0 \cdot 211423P) \qquad (2.69a)$$
$$P = (30/6) - (4/6)N. \qquad (2.69b)$$

Solving these two equations gives P of 1·70 and N of 4·96, the corresponding output being 45·16. At these input levels, the size of $p_y(MP_P)/p_P$ is 1·87. Since this ratio is greater than unity, the unconstrained profit maximizing solution involves an outlay in excess of 30. The constraint is therefore an effective one and the solution of equations (2.69) yields best operating conditions. The situation is illustrated graphically in Fig. 2.6 where the line AB is the isocline of equation (2.69a) and the line DE is the iso-cost locus of equation (2.69b). Their point of intersection, F, locates the optimal quantities of N and P.

r Response Processes, Each with Outlay Limited

For an array of r independent response processes each with an outlay restriction of the form

$$\sum_i p_i X_{ih} \leq C_h \qquad (h = 1, 2, \ldots, r) \qquad (2.70)$$

the set of overall best operating conditions is derived by applying to each individual response process the procedure outlined in the preceding section. The logic of this is that since each response process and its outlay constraint are independent of any other process or constraint, maximum overall profit can only occur if each process is at its own best operating conditions.

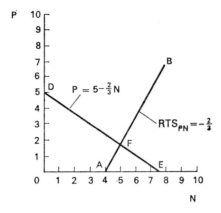

Fig. 2.6. Graphical solution of equations (2.69) to determine best operating conditions for response process (2.16) with limited total outlay.

r Response Processes, Overall Outlay Limited

The most important response constraint is that in which a group of independent response processes

$$Y_h = f_h(X_1, X_2, \ldots, X_n) \qquad (h = 1, 2, \ldots, r) \qquad (2.71)$$

have to be operated subject to the single outlay restriction:

$$\sum\sum p_i X_{ih} \leq C. \qquad (2.72)$$

The constrained objective function is

$$\pi = \sum p_h Y_h - \sum\sum p_i X_{ih} + \lambda(\sum\sum p_i X_{ih} - C). \qquad (2.73)$$

Setting $\partial \pi / \partial X_{ih}$ and $\partial \pi / \partial \lambda$ equal to zero, we have the $rn + 1$ equations

$$p_h(\partial Y_h/\partial X_i) - p_i + \lambda p_i = 0, \qquad (2.74a)$$

$$\sum\sum p_i X_{ih} - C = 0. \qquad (2.74b)$$

Eliminating λ from equations (2.74a) and rearranging equation (2.74b) as the iso-cost locus, we have the rn equations

$$(p_{y1}/p_1)(\partial Y_1/\partial X_1) = (p_h/p_i)(\partial Y_h/\partial X_i), \qquad (2.75a)$$

$$X_{11} = C/p_1 - \sum X_{1k} - \sum\sum (p_j/p_1) X_{jh}, \qquad (2.75b)$$

where $h = 1, 2, \ldots, r; k = 2, 3, \ldots, r; i = 1, 2, \ldots, n; j = 2, 3, \ldots, n;$ and $i \neq 1$ if $h = 1$. Simultaneous solution of these rn equations gives the r sets of n input levels for maximum profit with total outlay equal to C. As for the case of a single response process with outlay limited, should the ratio $p_h(MP_{ih})/p_i$ be less than unity, the outlay constraint is not effective and best operating conditions are the unconstrained ones. Also, as before, should the solution of equations (2.75) involve any X_i negative, boundary solutions involving the isocline segment $X_i = 0$ must be obtained.

Finally, notice that equations (2.75a) are identical to equations (2.52) and can be categorized into the $r - 1$ product–product relations, $n - 1$ factor–factor relations and $(n - 1)(r - 1)$ factor–product relations, respectively, of equations (2.54), (2.55) and (2.56). Indeed, with r independent processes, there is only one difference in procedure between maximizing profit subject to an overall outlay constraint and maximizing it subject to an overall product-value constraint. For an overall outlay constraint, the solution involves the intersection of isoclines with the iso-cost plane; for an overall product–value constraint, the solution is the intersection of isoclines with the iso-revenue plane.

Example

Suppose best operating conditions are required for response functions (2.57) and (2.58) when p_N is 6, p_P is 8, p_y is 10, and outlay is limited to 80.

The constrained objective function is

$$\pi = 10(Y_1 + Y_2) - 6(N_1 + N_2) - 8(P_1 + P_2)$$
$$+ \lambda(6(N_1 + N_2) + 8(P_1 + P_2) - 80). \quad (2.76)$$

Corresponding to equations (2.75), the profit-maximizing conditions are

$$(10/6)\,(\partial Y_1/\partial N_1) = (10/6)\,(\partial Y_2/\partial N_2), \quad (2.77a)$$

$$(10/6)\,(\partial Y_1/\partial N_1) = (10/6)\,(\partial Y_1/\partial P_1), \quad (2.77b)$$

$$(10/6)\,(\partial Y_1/\partial N_1) = (10/8)\,(\partial Y_2/\partial P_2), \quad (2.77c)$$

$$P_1 = 80/8 - P_2 - 6(N_1 + N_2)/8. \quad (2.77d)$$

Substituting for the marginal products from the parent response function and solving, these equations give constrained best operating conditions of $P_1 = 3\cdot00$, $P_2 = 1\cdot44$, $N_1 = 5\cdot54$ and $N_2 = 1\cdot87$ for an overall outlay of 80. Since the value of the ratio $p_y(MP_P)/p_P$ is greater than unity, the restraint is an effective one and unconstrained profit maximization is infeasible.

2.10 Further Reading

Extensive non-algebraic discussion of unconstrained best operating conditions is to be found in Heady (1952, chs. 4, 6, 7, 8). Full mathematical treatment of constrained and unconstrained production efficiency for both single and multiple product situations is given by Dano (1966) and by Frisch (1965), who considers both the situation with fixed prices and the situation where prices are functionally related to the scale of the production process. Henderson and Quandt (1971, ch. 3) present a rigorous mathematical outline of both constrained and unconstrained profit maximization. Agricultural discussion of constrained response is to be found in Tramel (1957), Doll (1959), Duloy (1959), Heady and Dillon (1961, ch. 2), Abraham (1965), Colwell and Esdaile (1966), and Anderson (1967a). The latter reference discusses the alternative procedure of meeting an outlay constraint by incorporating an opportunity cost element in the analysis as also does Colwell (1973a and 1976). The mechanics of using Lagrangian multipliers to solve constrained maximization problems is outlined in most intermediate and advanced level texts on differential calculus. Frisch (1965, ch. 10) discusses the economic interpretation of the Lagrangian multiplier in constrained response efficiency analysis. General applications of Lagrangian methods to response analysis are discussed by Draper (1963), and also by Myers and Carter (1973) and Umland and Smith (1959) who consider the situation where the constraint is itself a function of the input variables. For example, in milk production a relevant constraint might be the percentage of butterfat. Since butterfat content, like milk yield, is a function of the feed inputs, the relevant constraint in the milk objective function would involve the response function for butterfat. Anderson (1967a) discusses such analyses in a specifically agricultural context.

The general role of prices as market-oriented weights in the objective function pertinent to best operating conditions is discussed by Dorfman (1964, especially chs. 2 and 6), Frisch (1965, ch. 1) and, with special reference to crop-fertilizer response, by Anderson (1967a). The relationship between a production function and its associated cost functions—a topic we have not considered —is outlined by Tangri (1966) and Dano (1966). Likewise the determination of normative input demand and output supply functions from response functions is outlined by Heady and Dillon (1961, pp. 59–64) and exemplified by Heady and Tweeten (1963, ch. 6), Heady, Pesek and Rao (1966), Ogunfowora and Norman (1973) and Tweeten and Heady (1962). The influence of taxation on factor use is discussed by Candler and Cartwright (1970).

The fact that profit or gross margin is relatively insensitive in the region of best operating conditions to non-optimal levels of inputs has been variously commented upon by Heady and Pesek (1955), Hutton (1955), Hutton and Thorne (1955) and, more recently, by Anderson (1968a, 1975 and 1976), Colwell (1970 and 1976), Doll (1972), Englestad (1963), Godden and Helyar (1975), Havlicek and Seagraves (1962), Jardine (1975a and b) and Perrin (1976). The relative insensitivity of profit arises because the response function is generally smoothly rather than sharply curved, and because marginal profitability is thus necessarily close to zero in the region of best operating conditions. However, while gross margin may be *relatively* insensitive to errors in input use, the *absolute* cost of not having best operating conditions may still be significant. This will be especially so if large numbers of technical units (hectares or animals) are involved.

A large number of empirical studies of crop and livestock response efficiency analysis are collated and discussed in—to cite a few of many studies—Heady and Dillon (1961, chs. 8–15), Abraham and Rao (1960), Alcantara and Prato (1973), Baird and Mason (1959), Carley (1973), Carter, Dean and McCorkle (1960), Colwell (1973b), de Datta and Barker (1968), de Janvry (1972a), de Oliveira (1973), Davis, Sundquist and Frakes (1959), Dean et al. (1972), Dent et al. (1970), Dent, English and Raeburn (1970), Gastal (1971), Hoffnar and Johnson (1966), Hoover et al. (1967), Montero and Perez (1967), Russell (1968c), Ryan and Perrin (1973), Sharpe and Dent (1968), Sundquist and Robertson (1959), Townsley (1968), Valdés (1967), Yaron (1971), and in reports by the National Research Council of the U.S.A. (1961) and the O.E.C.D. (1962, 1964, 1965, 1968a and b, 1969a and b). These O.E.C.D. reports also discuss the question of co-operation between the various scientific disciplines interested in crop and livestock response analysis, as do Heady (1966, ch. 6), Anderson (1968b) and Wragg (1970).

Assessment of the response function approach in terms of basic biological research requirements and examples of other approaches to response have been given by Blaxter (1961, 1962a and b), Blaxter, Graham and Wainman (1956), Helyar and Godden (1976), Sandland and Jones (1975), Sharpe and Dent (1966) and Milhorn (1966). The consideration of fertilizer response by Anon. (1974a) and Perrin (1976) is noteworthy for their use of a factor by factor model (based on Liebig's (1855) law of the minimum) involving a linear response up to some yield plateau as discussed by Anderson and Nelson (1975), Boyd (1972), Boyd et al. (1970), Boyd, Tong Kwong Yuen and Needham (1976) and Swanson (1963), and developed operationally by Waugh, Cate and Nelson (1973). For livestock, Fawcett (1973) has also investigated a Liebig-based approach. See also Upton and Dalton (1976).

2.11 Exercises

2.11.1. Recompute some of the data of Table 2.1 using not response function (2.16) but the alternative response function estimates of Exercise 1.9.2.

2.11.2. For each of the alternative response function estimates of Exercise 1.9.2, what combination of N and P would be least cost for yields of 20 and 40 if p_N/p_y where 1·0 and p_P/p_y were 1·5? Compare with Fig. 2.3.

2.11.3. How should X_1 and X_2 be allocated between the two response processes

$$Y_1 = X_1^{·7}, \qquad Y_2 = X_1^{·2} X_2^{·8}$$

to best achieve a total revenue of 500 if p_{y1} is 10, p_{y2} is 15, p_1 is 1, and p_2 is 2? Illustrate your solution graphically in an analogous fashion to Figs. 2.4 and 2.5.

2.11.4. For each of the alternative response function estimates of exercise 1.9.2, what combination of N and P would be most profitable if total outlay were limited to 30; p_N being 4, p_P being 6, and p_y being 5. Compare with Fig. 2.6.

2.11.5. In an analogous fashion to Figs. 2.4 and 2.5, represent the solution of equations (2.77) diagrammatically.

2.11.6. How should X_1 and X_2 be allocated between the two response processes of Exercise 2.11.3 if total outlay were limited to 10 and prices were unchanged? What if all prices were doubled?

2.11.7. Assess the relative computational ease with which the response functions of Exercise 1.9.3 lend themselves to determining best operating conditions.

2.11.8. For simultaneous operation of the two response processes

$$Y_1 = a_0 X_{11}^{a_1} X_{21}^{a_2}, \quad Y_2 = b_0 X_{12}^{b_1} X_{22}^{b_2},$$

under the input quantity restrictions

$$X_{11} + X_{12} = k_1, \quad X_{21} + X_{22} = k_2,$$

derive the function $Y_2 = f(Y_1)$ specifying the locus of least-cost combinations of Y_1 and Y_2. Show that it reduces to

$$Y_2 = b_0 k_1^{b_1} k_2^{b_2}$$

if Y_1 is zero. Finally, using input prices of 1 for p_1 and p_2, and values of 100 for k_1 and k_2, 1 for a_0 and b_0, and 0.4 for b_1 and b_2, graph the least-cost locus for the three cases when a_1 and a_2 are both equal to (i) 0.2, (ii) 0.5, and (iii) 0.8.

2.11.9. For the fertilizer response function $Y = a + bX^{.5} + cX$, show that $\pi_{k0} = (2k^{.5} - k)\pi_0$, where π_0 is the fertilizer gross margin with X equal to its optimal level X_0, and π_{k0} is the gross margin when $X = kX_0$ ($k > 0$). Graph the ratio π_{k0}/π_0 for k from zero to two. What if $X^{.5} = Z$? What if there are other variable costs of size D? (*Hint*: See Jardine (1975b) if in trouble.)

2.11.10. By solving for λ in the various constrained situations of Section 2.9, deduce an interpretation of these λ's in terms of constrained response efficiency.

2.11.11. Derive best operating conditions for the response process

$$Y = f(X_1, X_2, \ldots, X_n)$$

subject to the outlay constraints

$$\sum p_i X_i = C_1 \quad (i = 1, 2, \ldots, m)$$
$$\sum p_j X_j = C_2 \quad (j = m+1, m+2, \ldots, n).$$

2.11.12. Determine best operating conditions for response functions (2.57) and (2.58) when p_N is 6, p_P is 8, p_y is 10 and the opportunity cost or desired return per unit of outlay is 10 per cent. What if total outlay is also limited to 80?

CHAPTER 3

Response Efficiency Over Time

3.1 Introduction

So far we have taken no account of time in relation to the analysis of crop or livestock response. We have treated response as if it were instantaneous or as if time were a fixed input. In fact, crop or livestock response processes are never instantaneous; and quite frequently, time is not a fixed input so that it has to be considered explicitly. Indeed, often the influence of time on response efficiency is much more pervasive and complex than the influence of physical inputs. Not only may time directly affect the physical response process, it may also influence response efficiency through time-price effects in the objective function or through uncertainty about the future. In this chapter we consider the role of time within the general framework of response efficiency under certainty of yields and prices, leaving considerations of risk till Chapter 4. As in prior chapters, specific crop or livestock response processes will only be used for illustrative purposes.

3.2 Time Influences on Response

There are four ways in which time may affect the physical response process. Firstly, the contribution of fixed inputs may vary with the time-length of the response process so that time directly influences response. In such cases time acts as a variable input which must be included explicitly in the response function. For example, the yield of pasture hay depends (among other things) on the time from the start of the growing season until harvest. In such cases, denoting time by t, we have the time-dependent response function:

$$Y = f(X_1, X_2, \ldots, X_n, t). \qquad (3.1)$$

Secondly, the capacity of the set of fixed inputs to accommodate variable inputs may be a function of time and of the mix of variable inputs made available. Thus the total consumption of feed by broilers depends on the time since hatching and the mix of carbohydrate and proteins in the ration. Algebraically, at one extreme we may have the situation where the utilization of each variable input depends only on time so that we have the set of n *input-utilization* equations

$$X_i = f_i(t) \qquad (3.2)$$

which, on substitution into the response function, gives output as a function of time:

$$Y = f(t). \qquad (3.3)$$

At the other extreme we may have the situation where the utilization of variable inputs over time can be specified by the single expression

$$X_1 = f(t, X_2, X_3, \ldots, X_n). \qquad (3.4)$$

Between these two extremes, input utilization over time may be specified by a set of less than n equations, some like equation (3.2) and some, for other inputs, of the form:

$$X_j = f_j(t, X_g, X_m, \ldots). \qquad (3.5)$$

Thirdly, the time sequence or pattern of either input injections or output harvests may influence yield. A sheep's wool production over a given period, for example, is influenced by any variations in its dietary pattern over the period. On the input side, if the sequence of input injections varies systematically over time and there is a single harvest, we have a set of n *input-supply* equations

$$X_i = f_i(t) \qquad (3.6)$$

where X_i is the cumulative input of X_i over the period t. Output, however, may depend not only on the level of X_i but also on the sequential pattern by which X_i has been injected into the response process. If so, the response function will be some function of equations (3.6) so that in some complicated fashion we again have:

$$Y = f(t). \qquad (3.7)$$

While equations (3.6) and (3.7) are somewhat analogous to equations

(3.2) and (3.3), the former relate to input supply while the latter refer to input demand.

Should harvesting of output occur either continuously over time or as a sequence of discrete harvests, cumulative yield must then also allow for the intensity of harvesting over time. For example, total milk production of a cow varies with the milking pattern followed.

Fourthly, input carry-over effects may occur from one production period to another if the injection of variable inputs within one production period is not completely utilized within that period. A common example is the partial carry-over of fertilizer inputs in pasture and crop production from one production period to another. In such cases, yield in the current production period will be a function of both current inputs and of the variable inputs carried over from prior periods, this carry-over being a function of time.

3.3 Time-Price Effects

Through their effect on response, the possible time influences outlined above also influence the objective function. Hence they also affect the choice of best operating conditions. As well, time may influence the objective function through time influences on prices and profit opportunities.

Firstly, input and output prices may vary over time in either a predictable or a subjectively assessable probabilistic fashion.

Secondly, the use of inputs over time in a particular response process means that these inputs (both fixed and variable) are tied up and are not available for use in some other process. Alternative uses of the inputs are forgone and an OPPORTUNITY COST is entailed. This input opportunity cost is the profit forgone from not using the physical inputs over time in their most profitable alternative use. So, although we talk of time opportunity cost, this cost depends not on time itself but on the availability of alternative uses for inputs over time. In broiler production, for example, the time opportunity cost of carrying a batch of broilers for another week (assuming there is no other more profitable use of inputs) is the contribution which this week could make to profit if it were used for a new batch of broilers.

Thirdly, there are interest rate or TIME PREFERENCE effects. In the sense in which we use the term time preference, these effects relate to such time-induced problems as the necessity: (a) to compound present costs or discount future returns so as to make them comparable; (b) to further discount future profits because of uncertainty on the basis that "A bird in the hand is worth two in the bush"; and (c) to use actuarial formulae to convert lump sums to flows over time, and vice versa.

All these time preference considerations involve the use in the analysis of some positive rate of interest per unit of time. If time preference is ignored, this is equivalent to assuming an interest rate of zero. Ideally, in a world of certainty, the interest rate used should be the maximum rate of return obtainable from using the inputs. This interest rate should be jointly determined (within a response process that is feasible for the inputs) along with the value to be attached to the fixed inputs used in the process. While these fixed inputs have some initial historical cost (such as land used in forestry), their relevant value for decisions over time depends not on this historical cost but on the profitability of the most profitable response process they can sustain. Of necessity, therefore, time preference analysis may become quite complicated, especially if—as in many response processes—input injections and output harvests are to be made in irregular sequences within a single run of the process.

3.4 Time and the Objective Function

In Chapter 2, ignoring time influences, we defined best operating conditions in terms of the input array which maximized a timeless objective function of the form: "profit" equals "output gains" minus "input losses". With the introduction of time, the efficiency indicator profit is still the difference between output gains and input losses. The algebraic form of the objective function, however, must accommodate the various time influences outlined above in Sections 3.2 and 3.3 should they be relevant.

In particular, since best operating conditions are desired over some chosen period of time into the future, if time preference is relevant, profit must be evaluated as the present value of the future sequence of profits. It is this present value or its equivalent in some form which must be maximized to achieve best operating conditions; for example, we may maximize the equivalent profit flow per unit of time over the planning period. Of course, if there is no time preference, present values and actually occurring values of profit will be identical.

Obviously, the time-dependent objective function may be quite complicated since real-world time influences may be much more complex than the simple outline given in Sections 3.2 and 3.3. As well, there is a further complication that may be relevant. Over a given response period it may be impossible to specify a single profit function encompassing all possible time sequences of input injections or output harvests. If so, a different profit function must be specified for each system. Best operating conditions then correspond to those conditions yielding maximum profit in whichever system of input injections and output harvests has the highest maximum profit.

Mathematically, the problem of ascertaining best operating conditions over a period of time may be approached via two procedures. One is that of the differential calculus whereby the overall problem is considered as a single problem in many variables. The other approach is that known as dynamic programming whereby the overall problem is approached as a series of recursively related problems each involving some few variables. Both approaches yield the same answers. Although dynamic programming may sometimes be the simpler approach, apart from using it to analyse the effects of input carry-over from one period to another and giving some references to it in the reading list at the end of this chapter, we will continue to analyse response efficiency via the differential calculus.

3.5 Planning over Time

There are various ways of planning best operating conditions over time. Which is the most efficient procedure will depend on the costs and benefits of the alternative procedures. At one extreme, best operating conditions may be determined once and for all at a given point in time for some fixed period into the future covering some number of response periods. At the other extreme, a continuous or rolling planning procedure may be followed with best operating conditions being evaluated continuously or at the start of each response period within the overall planning horizon. Both procedures allow for the overall future sequence of profits, and for the determination of the best time-length for each response period. If input and output prices are constant (as is assumed hereafter) and there is no time preference, both procedures will lead to the same set of best operating conditions; under other circumstances, the former procedure will generally lead to some degree of response inefficiency which must be balanced against its cheaper planning cost.

3.6 Unconstrained Profit Maximization over Time

In the timeless analysis of Chapter 2 we showed that unconstrained profit maximization occurred when each variable input was used at the level such that

$$p_y MP_i = p_i. \tag{3.8}$$

The most general statement of this timeless profit maximizing condition is that for best operating conditions the MARGINAL REVENUE or MARGINAL VALUE PRODUCT of X_i, given by $\partial(p_y Y)/\partial X_i$, must equal the MARGINAL COST of X_i, given by $\partial(p_i X_i)/\partial X_i$. In other words, the last increment of X_i must just pay for itself.

As we show below, exactly the same general principle of equating marginal revenue and marginal cost must prevail for profit maximization over time. The only difference is that marginal cost over time must allow for time opportunity cost and time preference effects as well as the direct marginal cost of inputs. Concomitantly, for best operating conditions over time, along with deciding on the level of X_i it is necessary to choose a time length of run for the response process.

Profit maximization over time involves more complex analysis than in the timeless case. To maintain as much simplicity as possible while still outlining the relevant principles, we first consider in Section 3.6.1 a simple response process operating over time but in the absence of time preference, i.e. a zero interest rate is assumed. This analysis spotlights the effect of the time opportunity cost of inputs. In Section 3.6.2 the analysis of this simple time-dependent process is extended to incorporate time preference effects. A numerical example illustrating the principles of Sections 3.6.1 and 3.6.2 is presented in Section 3.6.3. Throughout, input and output prices (p_i and p_y) are assumed to be constant over time, the question of price variability being completely ignored. It is also assumed that the response process operates on the basis of a given set of fixed inputs (e.g. buildings and equipment) about whose replacement we do not have to worry. In other words we are still only concerned with the best use of the variable inputs of Chapters 1 and 2. For most crop and livestock response processes this is not too severe a simplification. Ideally, however, the injection of capital items into a time process should enter into consideration, especially as the distinction between fixed and variable inputs vanishes if we take a long enough period for all inputs to become replaceable or variable.

3.6.1. WITHOUT TIME PREFERENCE

Consider a time-dependent response process

$$Y = f(X_1, X_2, \ldots, X_n) \tag{3.9}$$

$$X_i = f_i(t) \qquad (i = 1, 2, \ldots, n)) \tag{3.10}$$

with output being harvested at the end of the variable response period t. Inputs are supplied at the start of each response period and there are no input carry-over effects. The process is to be repeated continuously over time.

Assuming no input or output price changes over time, no constraints and no time preference, best operating conditions for each run of the response process must be identical; and maximization of total profit over a number of runs must be equivalent to maximizing profit per unit of time. Denoting profit per unit of time by π^* and the cost of the fixed inputs or the fixed set-up cost per response period by F, we have the unconstrained objective function:

$$\pi^* = (p_y Y - \sum p_i X_i - F)/t. \tag{3.11}$$

While the fixed input cost F is fixed for each run of the response process, the ratio F/t is not constant for t variable. Hence, unlike for the timeless analysis of Chapter 2, fixed input costs must be included in the time-dependent objective function.

Maximum profit per unit of time implies $\partial \pi^*/\partial t$ of equation (3.11) equal to zero. Thus, setting $\partial \pi^*/\partial t$ equal to zero, rearranging the equation, and remembering that $\partial F/\partial t$ is zero due to F being a constant, we must have

$$p_y(\partial Y/\partial t) - \sum p_i(\partial X_i/\partial t) = (p_y Y - \sum p_i X_i - F)/t \quad (3.12)$$

for maximum π^*. More simply if we denote by:

R: output gain or revenue per response period, $p_y Y$;

C: input loss or cost per response period, $\sum p_i X_i + F$;

then equation (3.12) can be written as

$$\partial R/\partial t - \partial C/\partial t = \pi^*. \quad (3.13)$$

In words, equations (3.12) and (3.13) imply that in the absence of time preference the marginal profit per unit of time (LHS) must equal the average profit per unit of time (RHS) if maximum profit per unit of time is to be achieved. This equality, moreover, implies that average profit per unit of time is at its maximum, as is easily proved. Since π^* equals $\pi^* t/t$, setting $d(\pi^* t/t)/dt$ equal to zero gives $d(\pi^* t)/dt$ equal to π/t as the condition for the average (π^*) to be maximized, i.e. the average is maximized when the marginal total profit equals average profit.

The profit maximizing criterion of equations (3.12) or (3.13) is illustrated in Fig. 3.1. Profit or net gain $(p_y Y - \sum p_i X_i - F)$ as a function of the length of the response period is shown by the curve OAB. Maximum profit from a single response period occurs at B with a response period of length OH. In contrast, maximum profit per unit of time occurs at A where the slope of the profit curve equals maximum average profit and the response period is only of length OG. Thus equation (3.13) is satisfied at A. As Fig. 3.1 indicates, best operating conditions for a sequence of response processes implies each response period should be shorter than if the response process was only to be carried through once. The logic of this is that as inputs are used beyond A, marginal profit per unit of time is less than the maximum average profit per unit of time that could be obtained by using these inputs in the next response period. Only if harvesting of output occurs at A of Fig. 3.1 will the marginal value product be equal to marginal cost over time.

In terms of the variable inputs, the condition for maximum profit per unit of time is that $\partial \pi^*/\partial X_i$ be zero. Thus, differentiating equation (3.11), we must have

$$p_y(\partial Y/\partial X_i) = p_i + (\partial t/\partial X_i)(p_y Y - \sum p_i X_i - F)/t \quad (3.14)$$

72 THE ANALYSIS OF RESPONSE IN CROP AND LIVESTOCK PRODUCTION

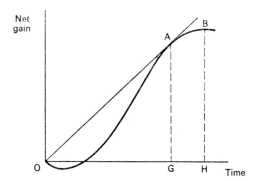

FIG. 3.1. Profit maximization over time in the absence of time preference.

or, akin to equation (3.13),

$$\partial R/\partial X_i = \partial C/\partial X_i + (\partial t/\partial X_i)\pi^*. \tag{3.15}$$

The LHS of equations (3.14) and (3.15) is the marginal value product of X_i; the expression on the RHS is the marginal cost of X_i. This marginal cost is the sum of two parts. The first of these is the cost of a unit of X_i without regard to time. We will call this the direct marginal cost of X_i. The second part is the time opportunity cost of a unit of X_i. It consists of the maximum average profit per unit of time possible in the next response period, $(p_y Y - \sum p_i X_i - F)/t$, multiplied by the time, $\partial t/\partial X_i$, required to utilize a unit of X_i.

Compared to the timeless analysis of Chapter 2, the effect of introducing time is to increase the marginal cost of X_i by the amount $(\partial t/\partial X_i)\pi^*$. In turn, this implies that best operating conditions over time involve lower levels of the variable inputs than are implied by the timeless analysis of Chapter 2. However, as discussed later, time opportunity-cost effects are irrelevant in certain response situations. In such cases the analysis of Chapter 2 remains appropriate.

Most importantly, relative to equation (3.11) it should be noted that no direct price was attached to time. Of itself, time has no price. None the less, our profit maximizing analysis shows how the availability of alternative uses for inputs over time introduces a time opportunity–cost element into the marginal cost of variable inputs.

While the above analysis has been couched in terms of the fairly simple case of a sequence of response periods based on equations (3.9) and (3.10), the same general principle applies in more complicated cases without time preference. Thus should physical relations or prices be different in the next response period, the rule still remains that response in the present period should be harvested when, per unit of time, its marginal net revenue is equal to the maximum average net revenue expected from the next response period. Concomitantly, variable inputs should be used to the point where their marginal revenue is equal to their overall marginal cost inclusive of time opportunity cost based on future revenue possibilities. The difficulty, of course, lies in making a good estimate of revenue possibilities from the next run of the response process.

3.6.2. WITH TIME PREFERENCE

What if the response process of equations (3.9) and (3.10) is to be repeated over time in the presence of time preference? It then becomes necessary: (a) to allow for the compounding of costs so as to make them comparable with future revenues; (b) to evaluate future profits in terms of present values; and (c) to apply actuarial formulae so as to convert lump-sum values to equivalent flows over time so that the differential calculus can be used. The procedure is as follows. Denote by:

- t the time length of the response period;
- s the number of response periods;
- r the time preference interest rate per unit of time; the present value of a unit of profit realized at time t in the future being $(1 + r)^{-t}$, and the compounded value at time t of a current unit of cost being $(1 + r)^t$;
- ρ the rate of interest which, under continuous compounding (or discounting), yields the same result as r. Thus ρ equals ln $(1 + r)$; and
- π the profit, $[p_y Y - (\sum p_i X_i + F)(1 + r)^t]$, realized at the end of each response period.

The stream of profits $\pi_1, \pi_2, \ldots, \pi_s$ realized at the end of the response intervals of length t constitute an annuity. Applying the standard formula for obtaining the present value of an annuity, the present value (P) of this sequence of profits is given by:

$$P = \pi[(1 + r)^{st} - 1]/\{[(1 + r)^t - 1](1 + r)^{st}\} \quad (3.16)$$

where $[(1 + r)^t - 1]$ is the interest rate per response period. P is a lump sum. Applying the amortization formula to obtain the equivalent flow of profit per unit of time over the period st, we have the unconstrained objective function:

$$\pi^{**} = P[\rho(1 + r)^{st}]/[(1 + r)^{st} - 1] \quad (3.17)$$

$$= \pi\rho/[(1 + r)^t - 1], \quad (3.18)$$

where π^{**} is the equivalent steady rate of profit flow per unit of time corresponding to a lump sum present value of P. Equivalently, we could have applied the continuous sinking-fund formula, $\rho/[(1 + r)^t - 1]$, direct to π to convert this lump-sum profit to its flow equivalent. Setting $\partial \pi^{**}/\partial t$ equal to zero, we obtain the following condition for maximum profit flow per unit of time:

$$(\partial \pi/\partial t)/(1 + r)^t = \pi^{**}. \quad (3.19)$$

Thus, with time preference, profit maximization occurs with that time length of response t such that at t the present value of the marginal profit per unit of time, $(\partial \pi/\partial t)/(1 + r)^t$, is equal to the flow of profit, π^{**}, which could be obtained by allocating that last unit of time to a new run of the response process.

If we denote by:

R: output gain or revenue, $p_y Y$, per response period,

C: cost, $(\sum p_i X_i + F)(1 + r)^t$, per response period,

so that π equals $R - C$, equation (3.19) can be expressed analogously to equation (3.13) as:

$$\partial R/\partial t - \partial C/\partial t = \pi^*(1 + r)^t \rho t/[(1 + r)^t - 1]. \quad (3.20)$$

For the usual case of r positive, the factor $(1 + r)^t \rho t/[(1 + r)^t - 1]$ is greater than unity. Hence, comparing equations (3.13) and (3.20), it follows that time preference implies a shorter optimal response period than in the absence of time preference. Logically enough, when r falls to its limit of zero so that there is no time-preference, equation (3.20) can be shown to collapse to equation (3.13).

An interesting rearrangement of equation (3.19) is obtained by expressing it in terms of the underlying response process. Thus we have

$$\partial[p_y Y - (1+r)^t(\sum p_i X_i + F)]/\partial t$$
$$= [p_y Y - (1+r)^t(\sum p_i X_i + F)](1+r)^t \rho/[(1+r)^t - 1] \quad (3.21)$$

which, on expansion and rearrangement, gives the condition for profit maximization as

$$\partial R/\partial t = R\rho + \pi^{**} + (1+r)^t \sum p_i (\partial X_i/\partial t). \quad (3.22)$$

Expressed in this form, the profit maximizing criterion is that the marginal revenue per unit of time (LHS) must equal the marginal cost per unit of time (RHS). This marginal cost is made up of three parts: $R\rho$ or $p_y Y \rho$ which is the return that could be earned during the current unit of time from the value of the product, $p_y Y$, if output were harvested now and this sum invested; π^{**} which is the opportunity cost suffered or profit foregone by not allocating the current time unit to a new run of the process; and $(1+r)^t \Sigma p_i (\partial X_i/\partial t)$ which is the compounded direct marginal cost of the variable inputs during the tth unit of time.

If (as is typically the case in forest analysis) input (regeneration) costs are regarded as independent of time, then equation (3.22) reduces to

$$\partial R/\partial t = R\rho + \pi^{**} \quad (3.23)$$

which is the classical Faustmann criterion for best operating conditions for long-period response processes.

Finally, since X_i in the response system of equations (3.9) and (3.10) is a function of t, from equation (3.20) we have

$$\partial R/\partial X_i = \partial C/\partial X_i + (\partial t/\partial X_i)\pi^*(1+r)^t \rho t/[(1+r)^t - 1] \quad (3.24)$$

as the requirement for best operating conditions with respect to X_i. Comparing equations (3.15) and (3.24), it is obvious that time preference increases the time opportunity cost of X_i and hence its overall marginal cost. As a result, time preference induces a reduction in the best operating level of each variable input.

76   THE ANALYSIS OF RESPONSE IN CROP AND LIVESTOCK PRODUCTION

Both our analyses with and without time preference have been based on a very simple prototype of time-dependent response. Empirical situations involving sequential response are likely to be far more complex, although simple models such as we have used may provide worthwhile approximations. Certainly it is mathematically straightforward to extend the analysis to encompass input injections made at any stage of the response period. Regardless of such possible extensions, the above analysis indicates the general rule for sequential best operating conditions with time preference: variable inputs must always be used at such level that their marginal revenue equals their marginal cost inclusive of time opportunity cost and time preference effects.

Except for lengthy response processes, such as in forest growth, or for long sequences of response, such as may be involved in pasture-development planning, time preference is generally of no great significance in crop or livestock response sequences. Accordingly, except for Sections 3.9.4. and 4.4.4. and the numerical example directly below, from here on we will ignore the possible role of time preference.

3.6.3. NUMERICAL EXAMPLE

As a simple numerical example of response analysis over time we use the time-dependent response process specified by the two equations:

$$Y = 10 + 100X_1 - X_1^2,$$
$$X_1 = t.$$

Prices of $p_y = 1$ and $p_1 = 1$, a fixed cost of $F = 910$ per response period, and an interest rate of 0·01, i.e. 1 per cent, per unit of time are assumed. Best operating conditions in terms of X_1 and t are required for continuous operation of the process over time with and without time preference. For production without regard to time, the principles of Chapter 2 indicate maximum profit occurs when X_1 is at a level of 49·5.

The relevant data corresponding to the criteria of equation (3.13) for the case without time preference and of equation (3.19) for the case with time preference are tabulated in Table 3.1 for increasing levels of input and output. This data indicates best-operating conditions occur for t and X_1 of 30 in the case without time preference, and for t and X_1 of approximately 27·18 in the case with time preference. To go beyond these input levels yields a marginal profit less than could be obtained by devoting the additional input of t or X_1 to a new run of the process. (The fact that t and X_1 are equal in this example is merely a consequence of the simple numerical form specified for the response process.) Note that while the empirical counterpart of equation (3.13) can be solved directly for optimal t, graphical procedures must be used to solve equation (3.19) because of its mixed exponential form.

3.7 Constrained Profit Maximization over Time

Should there be any constraints on the response sequence, constrained best operating conditions are found (as in the timeless analysis of Chapter 2) by appropriately inserting the constraints in the objective function.

For example, suppose the response process of equations (3.9) and (3.10) is to be run continuously with the constraint that $\sum p_i X_i$ must equal some constant k in each response period. This implies a constraint of $\sum p_i X_i/t$ equal to k/t per unit of time. Incorporating the constraint, objective function (3.11) becomes

$$\pi^* = (p_y Y - \sum p_i X_i - F)/t + \lambda(\sum p_i X_i - k)/t. \quad (3.25)$$

Differentiating with respect to X_i and λ, we have the $n + 1$ profit-maximizing conditions:

$$p_y(\partial Y/\partial X_i) - p_i - (\partial t/\partial X_i)(p_y Y - \sum p_i X_i - F)/t + \lambda p_i$$
$$\qquad - \lambda(\partial t/\partial X_i)(\sum p_i X_i - k)/t = 0, \quad (3.26a)$$
$$\sum p_i X_i - k = 0. \quad (3.26b)$$

Elimination of λ from equations (3.26a), along with equation (3.26b) rearranged as the iso-cost locus, gives the following n equations to be solved for $X_1, X_2, \ldots, X_n$:

$$MNR_1/MNR_j = p_1/p_j, \quad (3.27a)$$
$$X_1 = k/p_1 - \sum(p_j/p_1)X_j \quad (3.27b)$$

where $j = 2, 3, \ldots n$; and MNR_i is the marginal net revenue over time of a unit of X_i as given by:

$$MNR_i = p_y(\partial Y/\partial X_i) - p_i - (\partial t/\partial X_i)(p_y Y - \sum p_i X_i - F)/t. \quad (3.28)$$

TABLE 3.1
NUMERICAL EXAMPLE OF PROFIT MAXIMIZATION OVER TIME[a]

Length of response period	Output	Without time preference		With time preference	
		Marginal profit[b]	Time opportunity cost of marginal profit[c]	Marginal profit[d]	Time opportunity cost of marginal profit[e]
t	Y	$\partial \pi / \partial t$	π^*	$\partial \pi / \partial t$	$\pi^{**}(1+r)^t$
5	485	89	−86	79·33	−98·19
10	910	79	−1	68·73	−11·11
15	1285	69	24	58·10	15·22
20	1610	59	34	47·43	26·33
25	1885	49	38	36·73	31·14
27·18	1989·25	44·64	38·71	**32·08**	**32·08**
30	2110	**39**	**39**	25·98	32·63
35	2285	29	38·29	15·20	32·17
40	2410	19	36·50	4·37	30·26
45	2485	9	34	−6·51	27·44
49·50	2509·75	0	31·32	−16·90	23·95
50	2510	−1	31	−17·43	23·74

[a] Based on the time-dependent response process of Section 3.6.3.
[b] $\partial \pi / \partial t = \partial(p_y Y - \Sigma p_i X_i - F)/\partial t = 99 - 2t$.
[c] $\pi^* = \pi/t = (p_y Y - \Sigma p_i X_i - F)/t = 99 - t - 900/t$.
[d] $\partial \pi / \partial t = \partial[p_y Y - (1+r)^t(\Sigma p_i X_i + F)]/\partial t$
  $= 100 - 2t - (1 \cdot 01)^t[1 + (t + 910) \ln 1 \cdot 01]$.
[e] $\pi^{**}(1+r)^t = [p_y Y - (1+r)^t(\Sigma p_i X_i + F)](1+r)^t(\ln 1 \cdot 01)/[(1+r)^t - 1]$
  $= [10 + 100t - t^2 - (1 \cdot 01)^t(t + 910)](1 \cdot 01)^t(\ln 1 \cdot 01)/[(1 \cdot 01)^t - 1]$.

Equations (3.27) might be contrasted with equations (2.66) relating to timeless analysis of the same input constraint; and also with equation (3.14) which says that in the time-dependent unconstrained case, X_i should be used at such a level that MNR_i is equal to zero.

3.8 Time Classification of Response Processes

Crop and livestock response processes may be classified in terms of the various ways they involve input injections and output harvests over time. On the input side, the major input injection possibilities, with or without carry-over effects, are as follows:

A. *Single input injection* at the start of the process—for example, fertilizer in cereal production.
B. *Sequence of input injections* over the response period. With respect to input quantity and/or time between injections, these injections may be either:
   B.1. *Invariable*—for example, broilers on a fixed ration; or
   B.2. *Variable*—for example, irrigation water in crop production.

On the output side, the major time-of-harvest possibilities are for there to be:

C. *Single harvest* at the end of the response period. The time-length of this response period may be either:
   C.1. *Invariable*—for example, a crop with a fixed maturation period; or
   C.2. *Variable*—for example, the feeding period in broiler production.
D. *Multiple harvests* over the response period. In terms of the time between harvests and/or their intensity or severity, the sequence of harvests may be either:
   D.1. *Invariable*—for example, milk production under a regular twice-daily milking routine; or
   D.2. *Variable*—for example, harvesting of pasture hay.

Further time-based distinctions might be made between response processes on the basis, firstly, of their continuity and, secondly, in terms of whether or not they involve multiple stages. Some response processes are continuous. They continue over time with no natural break into separate response periods. An example is pasture production in some regions. In contrast, other response processes are quite discrete. They exhibit a distinct break between runs of the response process. Such is the case with cereal crops which, however, may involve fertilizer input carry-over effects.

Likewise, some processes involve only a single stage, whereas others may involve a number of stages. Thus cereal production is a single-stage process in contrast to wool production which involves the two concurrent stages of pasture production from the land and grazing of the pasture by the sheep to yield wool. In other processes, the multiple stages may be sequential rather than concurrent. Feedlot fattening of cattle, for example, may follow on from prior production of hay and grain.

These time-based distinctions between response processes could be extended much further. But simple as it is, the above classification scheme leads to 96 possible types of response processes on a time-oriented basis. For each of these 96 types, a mathematical model incorporating the response function, time relations, profit and opportunity cost considerations could be developed for formal analysis of best operating conditions. Such a possibility, of course, is the real advantage of organizing such a classification scheme. It immediately highlights the distinctive time-dependent characteristics that response processes may exhibit. However, rather than formally develop models for the various possible classes of response and analyse their implications, we will simply look at four specific response processes that, between them, straddle many of the time-dependent features mentioned above.

Before considering these examples, it should be noted that many crop-fertilizer response processes may be satisfactorily represented as having a fixed-response period with variable inputs injected at the start of the response period and a single output harvest at the end. Time plays no role in such an approximation, either as a variable in the response function or in terms of opportunity-cost effects in the objective function. As a result, best operating conditions for such processes may often be reasonably ascertained via the timeless principles of Chapter 2—a not insignificant fact given the importance of crop–fertilizer processes.

3.9 Examples of Time-dependent Response Analysis

In sequence, we will look at the determination of best operating conditions for: (i) a crop having a single input injection but multiple harvests; (ii) broiler production, which involves a sequence of feed injections, a single harvest and a variable response period; (iii) livestock production from pasture, a multi-stage process which may involve single or sequential input injections and single or multiple harvests; and (iv) a crop with fertilizer carry-over between response periods.

3.9.1. FERTILIZER FOR MULTI-HARVEST CROPS

As a relatively simple example of time-dependent response, we consider the case of crops such as berries, tomatoes, cantaloupes, melons, cucumbers, and many others which are characterized by:

(a) a single injection of fertilizer at the start of the response period;
(b) a sequence of harvests over the latter part of the response period as ripening of the crop proceeds;
(c) interaction between fertilizer inputs and the pattern of maturation or ripening of the crop.

We assume constant prices over time, no input carry-over effects, and harvesting of ripe fruit at constant intervals over the harvest period. Further assuming continuous repetition of the crop over time (as in glasshouse production of tomatoes), the decision variables for which best operating conditions are required are: (i) fertilizer input levels, (ii) number of harvests, and (iii) length of each run of the process. Appropriate choice of these variables hinges on modifying the analysis of Section 3.6.1 to allow for multiple harvests and the absence of input-utilization equations akin to equation (3.10).

Suppose for each harvest we have a yield function of the form:

$$y_t = b_0 + b_1 N + b_2 N^2 + b_3 P + b_4 P^2 + b_5 NP \\ + b_6 t + b_7 t^2 + b_8 Nt + b_9 Pt \quad (t = 1, 2, \ldots, k) \quad (3.29)$$

where y_t is output of the tth harvest or picking period, N is nitrogen input and P is phosphorus input. Diminishing returns are assumed to prevail to N, P and t. The terms in Nt and Pt allow for the influence of fertilizer on ripening, while the t and t^2 terms reflect the genetically determined growth pattern of the crop.

With some variable number of harvests k, total yield per run of the response process is†

$$\Upsilon = \sum y_t \quad (t = 1, 2, \ldots, k) \quad (3.30)$$
$$= k(b_0 + b_1 N + b_2 N^2 + b_3 P + b_4 P^2 + b_5 NP) \\ + (b_6 + b_8 N + b_9 P)(k^2 + k)/2 + b_7 k(k+1)(2k+1)/6. \quad (3.31)$$

† In equation (3.31) we make use of the fact that
$1 + 2 + 3 + \ldots + k = (k^2 + k)/2$
and that $1^2 + 2^2 + 3^2 + \ldots + k^2 = k(k+1)(2k+1)/6$.

82   THE ANALYSIS OF RESPONSE IN CROP AND LIVESTOCK PRODUCTION

Further assuming there is a fixed period T between planting and the first harvest, the objective function for profit per unit of time corresponding to equation (3.11) is

$$\pi^* = (p_y Y - p_N N - p_P P - F)/(T + k). \quad (3.32)$$

Setting $\partial \pi^*/\partial k$, $\partial \pi^*/\partial N$ and $\partial \pi^*/\partial P$ equal to zero, we have the set of three equations

$$p_y(\partial Y/\partial k) = \pi^*, \quad (3.33a)$$

$$p_y(\partial Y/\partial N) = p_N, \quad (3.33b)$$

$$p_y(\partial Y/\partial P) = p_P. \quad (3.33c)$$

Simultaneous solution of these equations gives the required levels of N and P, number of harvests k, and length of run $T + k$. Note that the best choice of k can involve the cessation of harvesting before all the crop has ripened. The logic of this is as discussed relative to Fig. 3.1.

Example

As a simple example, consider the single-fertilizer multi-harvest situation where, corresponding to equations (3.29) and (3.31), we have

$$y_t = 0.05N - 0.0001N^2 + 10t - 0.5t^2 + 0.001Nt, \quad (3.34a)$$

$$Y = k(0.05N - 0.0001N^2) + (10 + 0.001N)(k^2 + k)/2$$
$$- 0.5k(k+1)(2k+1)/6. \quad (3.34b)$$

What are the best operating conditions for N and k if set-up costs F are zero, T is 10, p_N is 0.1, and p_y is 1.0?

Corresponding to equations (3.33) we have

$$[0.05N - 0.0001N^2 + (10 + 0.001N)(k + 0.5)$$
$$- (6k^2 + 6k + 1)/12](10 + k) = k(0.05N - 0.0001N^2)$$
$$+ (5 + 0.0005N)(k^2 + k) - (2k^3 + 3k^2 + k)/12 - 0.1N$$
$$\quad (3.35a)$$

$$k(0.05 - 0.0002N) + 0.0005(k^2 + k) = 0.1. \quad (3.35b)$$

Eliminating N from these two equations, we have

$$0.3343k^5 + 0.1073k^4 - 92.7583k^3$$
$$+ 86.5062k^2 + 50k + 250 = 0. \quad (3.36)$$

Using graphical procedures, solution of this polynomial indicates two possibly relevant solutions. These are for k equal to 1·25 or 28·20. For k equal to 1·25, N and Y are negative so this solution is infeasible and k equal to 28 is the relevant number of harvests. The corresponding level of N, based on equation (3.35b), is 254·64, implying a total yield of 481·32 units and π^* of 11·99.

3.9.2. FEEDING PERIOD AND RATIONS FOR BROILERS

In the most general sense, broiler production is a prototype of a class of problems known as "replacement problems". Indeed, with a little ingenuity any response process can be cast as a replacement problem. The essence of such problems is that a decision has to be made as to when some element of the production process should be replaced; for example: sheep in wool production, hens in egg production, seed and fertilizer in crop production, cows in dairy production, rams and ewes in fat-lamb production, pasture in grazing production. In a less general sense, broiler production is typical of the class of non-grazing livestock processes that are repeated continuously over time and are characterized by a single harvest at some variable time following a sequence of feed input injections whose composition is chosen by the producer. For such processes, as in broiler production, the decision variables for which best operating conditions are required are the length of the response period or time of replacement, the specification of the feed ration, and the number of animals to be fed. It is these decision variables that are our concern here.

For broiler production, as for other livestock response processes, a variety of models of the production process have been suggested. These differences in approach correspond to various interpretations of the decision-making power of animals in deciding how much of what feed they will eat. As well, alternatives exist in analysing response in terms of either feed types (maize, fishmeal, etc.) or basic feed constituents (digestible protein, carbohydrate, etc.) in the ration. Some of these alternative approaches are referred to in this chapter's suggestions for further reading.

Here, in the belief that it is the most logical approach, we formulate and analyse a broiler model based on the decision variables:

P: per cent protein in the ration;
C: calories per unit weight of feed;
t: length of the response period.

Both output (body weight per bird) and feed consumption (weight of ration consumed) are specified as functions of P, C and t. Thus for the response function we have

$$Y = f_y(P, C, t) \qquad (3.37)$$

and for the utilization of feed

$$F = f_f(P, C, t) \qquad (3.38)$$

where Y is body weight per bird at harvest time t, and F is total weight of feed consumed up till harvest at time t.

If the cost (purchase, debeaking, medication) of each chick is K_1, profit per bird (π_b) with a response period of t is

$$\pi_b = p_y Y - p_F F - K_1 \qquad (3.39)$$

where p_y and p_F are the price per weight unit of Y and F respectively. While p_y (regarded as constant) is given from outside the response process, p_F is taken as a linear function of P and C, as given by

$$p_F = p_0 + p_P P + p_C C \qquad (3.40)$$

where p_0 is a constant, p_P is the cost of adding an additional 1 per cent of protein per unit of F, and p_C is the cost of adding an additional calorie to one unit of F.

The need for equation (3.40) arises from the fact that prices are usually known in terms of the feed components (such as maize and fishmeal) but not in terms of protein and calories. Of course, any ration specified in terms of protein and calories should be made up as the least-cost mixture from available sources of protein and calories.†

---

† The determination of such least-cost rations to meet given protein and calorie (and other) specifications is a problem in linear programming. See Heady and Candler (1958, pp. 131–45) and, for direct applications to livestock feed response, Brown and Arscott (1960), Dean *et al.* (1972), Dent (1964), Nelson and Castle (1958) and Nelson, Castle and Brown (1957).

Since broiler production is usually carried out with a fixed set of buildings, equipment and labour, the aim must be to maximize profit per unit of time with respect to this fixed set of production factors. A reasonable measure of these fixed factors is the amount of floor space available. Accordingly we express the objective function in terms of profit per unit of floor space per unit of time. To do so, we have to allow for the fact that space demands per bird are not constant. They increase over time, recommended space requirements apparently being met by a function of the form

$$Z = (c_0 + c_1 t + c_2 t^2)^{-1}, \qquad (3.41)$$

where Z is the number of birds of age t which can be accommodated by a unit of floor space (Z^{-1} being floor space per bird at age t), and c_0, c_1 and c_2 are estimated coefficients based on the relevant standard floor-space recommendations.

With fixed costs independent of flock size (such as labour, depreciation, repairs, etc.) of K_2 per unit of floor space per time unit, and a break of T time units between runs of the response process, we have the objective function

$$\pi^* = [Z(p_y Y - p_F F - K_1)/(T + t)] - K_2, \qquad (3.42)$$

where π^* is profit per unit of floor space per unit of time. Note that equation (3.42) is merely a modification of equation (3.11) to allow for the specific features of the broiler response process. The only variables in the objective function are Z, Y, F, p_F and t. Of these, Y and F are functions of P, C and t; p_F is a function of P and C; and Z is a function of t. Hence maximization of π^* implies setting the derivatives $\partial \pi^*/\partial t$, $\partial \pi^*/\partial P$ and $\partial \pi^*/\partial C$ equal to zero and solving simultaneously for P, C and t. The three equations to be solved are as follows:

$$p_y(\partial Y/\partial t) = p_F(\partial F/\partial t) + \pi_b/(T + t) + \pi_b Z(c_1 + 2c_2 t), \qquad (3.43a)$$

$$p_y(\partial Y/\partial P) = p_F(\partial F/\partial P) + p_P F, \qquad (3.43b)$$

$$p_y(\partial Y/\partial C) = p_F(\partial F/\partial C) + p_C F. \qquad (3.43c)$$

Having solved these equations to determine the best level of P, C and t, substitution of this t value into equation (3.41) gives the number of birds to be purchased per unit of available floor space for each run of the process.

In terms of specifying best-operating conditions, the above model (in common with those that others have postulated) has three major weaknesses. First, it specifies only a single ration for the entire response period and makes no allowance for changes in the protein or calorie level of the ration over time. In fact, ideal best operating conditions would likely involve a changing proportion of protein to calories in the ration as the birds approach harvesting. A model encompassing this possibility would involve a series of equations specifying body weight and feed consumption at the end of each week of the response period. With additional modifications, the model could then be set up so that the decision variables were the quantity of protein and calories to be included in the ration for each week. The second weakness of the model is that it ignores restraints on the broiler process arising from its typical role as an intermediate stage in a vertically integrated chain of processes. Thirdly, because it assumes non-overlapping batches of broilers with batch size determined by the floor-space requirement per bird at harvest, the model involves under-utilization of space. This waste of floor space, which can have a significant effect upon possible profit, gradually declines to zero at harvest. An extension of the model to eliminate this waste of space by allowing for the staggering or overlapping of response periods is noted in this chapter's further reading suggestions.

3.9.3. LIVESTOCK PRODUCTION FROM PASTURE GRAZING

Five major complexities must be allowed for in any reasonable model of the pasture-livestock complex.

First, there is the possibility of innumerable systems of grazing arising from the array of possible combinations of various time sequences of input injections and output harvests. Each system will have its own response function. The problem of best operating conditions, therefore, is not simply to decide on the level of input and output. It is also necessary to choose between alternative grazing systems.

Second, there is the multi-stage nature of the overall production process. Pasture production constitutes one stage, and its grazing for livestock production the next stage.

Third, these two production stages are not independent. Livestock influence pasture output; pasture output influences livestock production.

Hence the two stages interact over time so that allowance must be made for their simultaneous determination.

Fourth, allowance must be made for the possibility of conserving some pasture as hay or silage for later injection into the livestock production process.

Fifth, livestock make decisions and have variable appetites. Within limits, it is they (not the grazier) who decide how much of what feed will be eaten.

Given all these complicating features of the livestock grazing process, a full realistic model would be exceedingly complex. In contrast, only a simple model is presented here. The basic approach to best operating conditions is to balance scarce pasture feed supplies against revenue-producing demands for feed, all within the constraints imposed by livestock maintenance requirements and the cyclical nature of pasture growth and deterioration.

Assumptions

We distort the real world by assuming that:

1. The grazier operates on a fixed land area of uniform soil type devoted solely to established pasture of uniform character and history.

2. Newly produced pasture is of uniform nutritional quality regardless of its production date.

3. The yearly pattern of pasture growth is associated perfectly with the known climatic cycle.

4. A single class of livestock and a single fertilizer are the only variable inputs used in pasture production, fertilizer being applied only at the start of the yearly production cycle.

5. Only a single type of conserved pasture is produced and only a single type of supplementary feed is purchased.

6. No agistment or conserved fodder is sold.

7. There are no mechanical losses in conserving pasture or distributing supplementary feed; and storage costs of conserved pasture and purchased supplement are zero.

8. The rate of stocking in terms of the number of livestock is constant over land and time.

9. Initially, only a single grazing system is to be considered. This system is specified by some given time sequence of supplementary feeding and a given time sequence of minimal livestock liveweights (or some other criterion of the standard of livestock maintenance).

10. Prices do not vary over time and profit is to be maximized over the 12 months of the annual pasture cycle, the initial time point of each year-long response period being the beginning of the pasture flush.

11. Time preference is irrelevant, a unit of money here and now being worth no more (or less) than a unit at some future date.

Notation

Throughout, a dot over a variable is used to denote its time rate of change. Time subscripts are used in the form X_t to indicate the size of X at time t; and in the form $X_{0,t}$ to indicate the size of $(X_t - X_0)$. Where necessary, an asterisk is used to distinguish levels of a variable decided by the grazier rather than by the livestock. In order of occurrence, the more important symbols used are as follows:

R^* minimum allowable liveweight of an animal,
t time (in weeks, $0 \leq t \leq 52$),
B^* quantity of purchased feed fed out,
P quantity of pasture produced,
S number of livestock grazed,
F input of fertilizer,
H quantity of pasture conserved,
D quantity of pasture lost by deterioration,
C quantity of pasture consumed,
B quantity of purchased feed consumed,
E quantity of conserved pasture consumed,
Y quantity of livestock product,
R liveweight of livestock,
π profit per annual pasture–livestock cycle,
p_i price or cost per unit of the ith factor or product,
E^* quantity of conserved pasture fed out.

Grazing Model

For a given grazing system, specified by the preselected time sequences of minimal livestock liveweight (R^*_t) and purchased feed input $(B^*_{0,t})$, the pasture–livestock response process is depicted by an interdependent system of seven differential equations. As well, there are four restrictions (apart from the obvious non-negativity requirements for input and output) that ensure feasibility of the grazing system within the bounds specified by the mechanics of grazing production. Infallibility is not

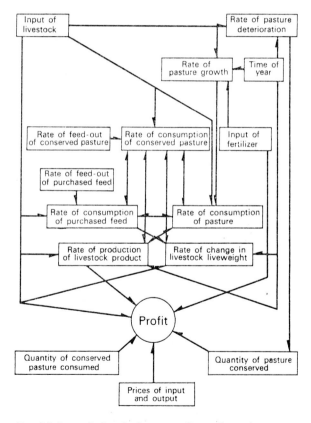

FIG. 3.2. Interrelations in the pasture-livestock grazing process.

claimed for the choice of variables entering the various relations. As yet, too little research has been done on the interrelationships of the grazing complex to specify anyways accurately all the relationships involved. Rather, the model must be regarded as a simple sketch of the way the pasture–livestock complex may operate.

The major interrelations or directions of influence allowed for by the model are depicted in Fig. 3.2. The specific relations of the model are as follows.

The rate of pasture production is specified as:

$$\dot{P} = f_1(S, F, t). \tag{3.44}$$

Although the input of livestock and fertilizer in prior pasture cycles undoubtedly influence the rate of pasture production, here they are taken as given. Also, a more realistic model would need to include the quantity of pasture conserved $(H_{0, t})$ in f_1. This would necessitate further circumscribing the grazing system by a given $H_{0, t}$ sequence.

The rate of pasture deterioration is specified as

$$\dot{D} = f_2(S, t) \tag{3.45}$$

and is assumed to be measured in such a way that remaining pasture can be assessed in constant units (say on an absolute protein basis or some such). It might also be argued that f_2 should include the stock of pasture on hand after livestock and mower have had their fill.

The rate of feed consumption by livestock is specified as a set of three equations, one each for pasture, purchased feed and conserved pasture. The equations are:

$$\dot{C} = f_3(S, \dot{P}, \dot{B}, \dot{E}, \dot{Y}, \dot{R}), \tag{3.46}$$

$$\dot{B} = f_4(S, \dot{C}, \dot{E}, \dot{Y}, \dot{R}), \tag{3.47}$$

$$\dot{E} = f_5(S, \dot{C}, \dot{B}, \dot{Y}, \dot{R}). \tag{3.48}$$

These feed consumption relations reflect the decision making capabilities of livestock should they ever be confronted with all three types of feed at the same time.

The rates of livestock production and maintenance are specified by the equations

$$\dot{Y} = f_6(S, \dot{C}, \dot{B}, \dot{E}, \dot{R}, t), \tag{3.49}$$

$$\dot{R} = f_7(S, \dot{C}, \dot{B}, \dot{E}, \dot{Y}, t). \tag{3.50}$$

If animal product and liveweight are synonymous, equations (3.49) and (3.50) collapse to a single equation.

Denoting the fixed cost per annual run of the pasture cycle by K, and assuming no variable cost in the harvesting of livestock product and no set-up cost for pasture conservation, the objective function for the pre-specified system of grazing is given by

$$\pi = p_y Y_{0,52} - p_S S - p_F F - p_H H_{0,52} - p_E E^*_{0,52} - K. \quad (3.51)$$

Apart from the usual non-negativity requirements for inputs and outputs, there are four other restrictions that must be met. The first of these relates to livestock maintenance. Using liveweight as a maintenance criterion, this restraint can be expressed as the necessity for the actual weight of livestock at any time-point (R_t) to equal or exceed the pre-specified minimal allowable liveweight at that same time-point (R^*_t). This minimal liveweight may or may not vary over the production cycle since it depends on the grazier's discretion. Thus we require

$$R_t \geq R^*_t. \quad (3.52)$$

The second restriction accommodates the necessity for the quantity of purchased feed consumed $(B_{0,t})$ to equal or be less than the quantity of purchased feed fed out $(B^*_{0,t})$ over the period from 0 to t. Hence we require

$$B^*_{0,t} \geq B_{0,t}. \quad (3.53)$$

The remaining two restrictions, detailed in equation (3.54), ensure that only "excess" pasture is conserved and that the consumption of conserved pasture does not exceed its supply at any stage of the consumption cycle. Denoting the reciprocal of the transformation coefficient between pasture and conserved pasture by k, we require:

$$P_{0,t} - D_{0,t} - C_{0,t} \geq H_{0,t} \geq k E_{0,t}. \quad (3.54)$$

Except for the maintenance requirement, all the above restrictions would be automatically satisfied in the real world. This would not necessarily be the case, however, in predictive or normative manipulation of the response model. Hence the necessity for precise specification of the restrictions.

Best Operating Conditions

Maximum profit under the given system of grazing (i.e. for the given time sequences of minimal liveweights and injections of purchased feed) implies maximization of equation (3.51) subject to the restrictions of equations (3.52), (3.53) and (3.54). Assuming no carry-over of conserved pasture, profit maximization implies all pasture conserved is consumed so that

$$k^{-1}H_{0.52} = E^*_{0.52} = E_{0.52}. \qquad (3.55)$$

Hence the objective function (3.51) can be written as:

$$\pi = p_y Y_{0.52} - p_s S - p_F F - (p_H k + p_E) E_{0.52} - K. \qquad (3.56)$$

Within the grazing livestock response process depicted by equations (3.44) to (3.50), the only variables under the grazier's control are S, F and t. However, t is not a relevant decision variable for objective function (3.56) since it has been removed by integration to obtain $Y_{0.52}$ and $E_{0.52}$. Accordingly, making use of relations (3.44) to (3.50), π can be expressed as a function of S and F alone. Maximization of π with respect to S and F, subject to the constraints (3.52), (3.53) and (3.54), gives the optimal level of livestock and fertilizer for the given grazing system. The optimal amount of conserved pasture is then obtainable from equation (3.48) expressed in its reduced form as a function of S and F. A harvest sequence to obtain this quantity of conserved pasture could then be deduced from the amount of excess pasture available over the pasture cycle, as defined by restriction (3.54) with livestock and fertilizer each at their optimal level.

Still, there would be no guarantee that this grazing system would be efficient. Efficiency will only prevail if no purchased feed fed out is wasted. If any purchased feed is not consumed, fixed costs could be reduced without decreasing production. However, once optimal livestock and fertilizer inputs have been determined for the original grazing system, an efficient system could be deduced by nominating, via the reduced form of equation (3.47), a sequence of purchased feed offerings that allowed no waste.

So much for a single grazing system. In fact, the model is relevant to an infinite number of grazing systems corresponding to all possible combinations of time sequences of purchased feed offerings and minimal livestock liveweights. To specify the optimal system, it would be necessary to compare the profitability of each efficient system. Perhaps the best way of doing this would be to investigate a range of grazing systems, look at the pattern of profit behaviour and thereby approach a satisfactorily "best" system. The problem, of course, would be immeasurably more difficult for a more realistic model involving a wider array of inputs, carry-over effects between years, variable stocking rates over land and time, diverse pasture types and locations, and probabilistic climatic elements.† Simulation procedures could perhaps be used but they would still necessitate knowledge of the important pasture–livestock response parameters.

3.9.4. CROP PRODUCTION WITH FERTILIZER CARRY-OVER

As a simple example of the determination of best operating conditions when there is carry-over of variable inputs between periods or runs of the response process, we consider the case of cereal crop production with fertilizer carry-over. The procedure used is that of dynamic programming. For simplicity, we examine the case of only a single fertilizer injected at the start of each run of the process. The analysis is easily extended to more than one type of fertilizer.

We suppose that the amount of carry-over to the beginning of period T from Q units of fertilizer available at the beginning of period 1 is $QV_1V_2 \ldots V_{T-1}$ with $0 \leq V_t \leq 1$ for each period t ($t = 1, 2, \ldots, T-1$) corresponding to a prior run of the crop response process. The carry-over coefficient V_t would depend on such factors as weather conditions and yield during period t.

As in the usual reverse time-ordering of dynamic programming, we use n to denote that period after which $n - 1$ further runs of the response process are to be made or equivalently, at the start of period n, there are n runs of the process still to be made. The response function $\Upsilon_n\{Q_n\}$ of

---

† In fact, even if mechanical procedures were available to obtain the necessary experimental observations on the variables of equations (3.44) to (3.50), these equations could not be estimated because the system contains too many variables that are determined within the system, i.e. the system is under-identified. (See Heady and Dillon, 1961, p. 139.) Compromise procedures for overcoming this difficulty are discussed in Dillon and Burley (1961, pp. 129–32).

grain yield in period n, denoted Y_n, to total available fertilizer at the start of period n, denoted Q_n, is assumed to exhibit diminishing returns so that the required second-order conditions for optimality hold true. Available fertilizer, Q_n, consists of residual or carry-over fertilizer, R_n, already in the soil at the beginning of period n, plus the fertilizer applied in period n which, of course, equals $(Q_n - R_n)$. The unit prices of grain and fertilizer in period n are denoted p_{yn} and p_{fn} respectively, and $\alpha(=1/(1+r))$ of Section 3.6.2) is used to denote the time preference discount factor per period.

Abstracting from any uncertainties in price or yield, recurrence equations of the usual dynamic programming form for finding the optimal application of fertilizer may be formulated. For the case with only one period remaining, i.e. $n = 1$, we have:

$$f_1\{R_1\} = \overset{\max}{Q_1} \; [\alpha p_{y1} Y_1 \{Q_1\} - p_{f1}(Q_1 - R_1)]; \qquad (3.57)$$

and for the general case with n periods remaining,

$$f_n\{R_n\} = \overset{\max}{Q_n} \; [\alpha p_{yn} Y_n \{Q_n\} - p_{fn}(Q_n - R_n) + \alpha f_{n-1}\{V_n Q_n\}], \quad (3.58)$$

where $f_n\{R_n\}$ is the return from following an optimal fertilizer policy in each of the n future periods given that residual fertilizer at the start of the current period, i.e. period n, is R_n. The cost of fertilizer is assumed to fall due at the beginning of each response period and revenue from the time-invariant single-harvest crop at the end of each period.

Differentiating the expression in square brackets in equation (3.57) with respect to our decision variable Q_1, and assuming the required second-order conditions hold (as implied by diminishing returns), the usual single-period condition for profit maximization with an opportunity cost of r per cent—as per equation (2.67c)—is obtained as

$$dY_1/dQ_1 = p_{f1}/\alpha p_{y1}. \qquad (3.59)$$

Defining the Q_1 which satisfies equation (3.59) as Q^*_1, equation (3.57) may be written as

$$f_1\{R_1\} = \alpha p_{y1} Y_1\{Q^*_1\} - p_{f1}(Q^*_1 - R_1), \qquad (3.60)$$

where the optimal final period application of fertilizer is $(Q^*_1 - R_1)$.

RESPONSE EFFICIENCY OVER TIME 95

Continuing for the case with two periods remaining, i.e. $n = 2$, from equation (3.58) we have

$$f_2\{R_2\} = \overset{\max}{Q_2} [\alpha p_{y2} Y_2\{Q_2\} - p_{f2}(Q_2 - R_2) \\ + \alpha[\alpha p_{y1} Y_1\{Q^*_1\} - p_{f1}(Q^*_1 - V_2 Q_2)]]. \quad (3.61)$$

Differentiating the expression within the outside square brackets with respect to Q_2 (the decision variable for $n = 2$) and assuming second-order conditions hold, the condition for profit maximization is found to be

$$dY_2/dQ_2 = (p_{f2} - \alpha V_2 p_{f1})/\alpha p_{y2}. \quad (3.62)$$

With Q^*_2 defined as the Q_2 satisfying equation (3.62), equation (3.61) becomes

$$f_2\{R_2\} = \alpha p_{y2} Y_2\{Q^*_2\} - p_{f2}(Q^*_2 - R_2) \\ + \alpha[\alpha p_{y1} Y_1\{Q^*_1\} - p_{f1}(Q^*_1 - V_2 Q^*_2)]. \quad (3.63)$$

Using the same procedure for the case with three periods remaining, the profit maximizing condition is

$$dY_3/dQ_3 = (p_{f3} - \alpha V_3 p_{f2})/\alpha p_{y3} \quad (3.64)$$

and we have

$$f_3\{R_3\} = \alpha p_{y3} Y_3\{Q^*_3\} - p_{f3}(Q^*_3 - R_3) \\ + \alpha[\alpha p_{y2} Y_2\{Q^*_2\} - p_{f2}(Q^*_2 - V_3 Q^*_3) \\ + \alpha[\alpha p_{y1} Y_1\{Q^*_1\} - p_{f1}(Q^*_1 - V_2 Q^*_2)]]. \quad (3.65)$$

Arguing by induction, these results may be extended to the general case with n periods remaining. Thus the optimal application of fertilizer in any period n is $(Q^*_n - R_n)$, where Q^*_n is such that

$$dY_n/dQ_n = (p_{fn} - \alpha V_n p_{fn-1})/\alpha p_{yn}. \quad (3.66)$$

The return from following such an optimal policy is given by

$$f_n\{R_n\} = \alpha p_{yn} Y_n\{Q^*_n\} - p_{fn}(Q^*_n - R_n) \\ + \sum_{i=1}^{n-1} \alpha^{n-i}[\alpha p_{yi} Y_i\{Q^*_i\} - p_{fi}(Q^*_i - V_{i+1} Q^*_{i+1})]. \quad (3.67)$$

Equation (3.66) shows that for determining optimal levels of fertilizer application taking account of carry-over between periods, the only relevant variable pertaining to future periods is the price of fertilizer in the next period. The response functions and grain prices that hold in future periods, being assumed independent of current fertilizer use, do not influence the current decision. Whatever the value of R_n, the optimal application of fertilizer is that amount which increases total fertilizer available to Q^*_n. As would be expected, the condition specified in equation (3.66) indicates that the greater the discount factor (i.e. the smaller the time preference interest rate) and the greater the rate of carry-over and the price of fertilizer next period, the greater should be the current application of fertilizer.

The above results relate to situations where the residual fertilizer or initial fertility at the beginning of a period is less than the optimal quantity of fertilizer for that period, i.e. $R_n < Q^*_n$. This will generally be true with continuous cropping. If $R_n \geq Q^*_n$, no fertilizer should be applied. This would be the case for fallow periods. In general, if k periods of fallow are to follow the crop produced in period n before another crop is grown, then the decision criterion of equation (3.66) becomes

$$dY_n/dQ_n = (p_{fn} - \alpha^{k+1} V_n V_{n-1} \cdots V_{n-k} p_{fn-k-1})/\alpha p_{yn}. \qquad (3.68)$$

3.10 Further Reading

The general principles of profit maximization over time are discussed at length in Winder and Trant (1961) and in the associated contributions by Faris (1960a and 1961). More formal treatment, but with the length of each response period fixed, is given by Henderson and Quandt (1971, pp. 309–33). Gaffney (1960), along with an excellent discussion of the Faustmann criterion—see also Gane (1968), has presented an excellent critique of many of the incorrect criteria that have been suggested. Although he deduces an incorrect criterion, the discussion by Boulding (1958) provides an interesting outline of the possible complexities in profit maximization over time. Frisch (1965, Pt. 5) presents a more formal analysis oriented to fixed input replacement which is of little relevance to crop and livestock response although his introductory discussion of time (ch. 4) is of interest. More relevant is the analysis of agricultural fixed asset replacement by Perrin (1972). Tax effects have been analysed by Chisholm (1974 and 1975) and Gaffney (1967 and 1970–1).

The standard formulae for obtaining the present value of an annuity and for amortizing a lump sum, as used in equations (3.16) and (3.17), are illustrated in Faris (1960a) and Chisholm and Dillon (1966).

An introductory outline of the alternative mathematical approach of dynamic programming is given in Sasieni, Yaspan and Friedman (1959, ch. 10), Roberts (1964) and Throsby (1964a and b). Using dynamic programming, the general problem of economic replacement has been outlined by Burt (1963). Among others, Burt and Stauber (1971), Dudley, Howell and Musgrave (1971), Flinn and Musgrave (1967), Hall and Buras (1961) and Hall, Asce and Butcher (1968) have used dynamic programming for economic appraisal and manipulation of crop response to irrigation water. Other response-oriented applications of dynamic programming have been by Kennedy (1972 and 1973) to beef feedlot control, by Hochman and Lee (1972) and Kennedy *et al.* (1976) to broiler production, and by Kennedy *et al.* (1973) and Stauber, Burt and Linse (1975) to optimal fertilizer use with carry-over between periods.

With the exceptions noted below, time opportunity–cost effects have largely been ignored in analyses of time-dependent crop and livestock response processes. For example, the livestock studies of Heady and Dillon (1961, chs. 8–13), while incorporating some consideration of the time required to achieve specified yields and comparisons of least-cost and least-time rations, do not really comprehend the problem of profit maximization over time. They fail to allow for the fact that animal production typically involves a continuous sequence of response periods so that best operating conditions necessitate a choice of length for each response period as well as a choice of feeds. The gradual development of a correct approach is illustrated, in sequence, by the broiler studies of Heady and Dillon (1961, ch. 10), Hansen and Mighell (1956), Brown and Arscott (1970), E. Smith (1965), and Trant and Winder (1961). A comparison of alternative approaches to the analysis of broiler response has been made by Hoepner and Freund (1964) whose analysis largely forms the basis of Section 3.9.2. This analysis has been further extended by Hadar (1965) to allow for the full utilization of floor space by concurrently running batches of broilers at various stages of development, and by Hochman and Lee (1972) and Kennedy et al. (1976) using a dynamic programming approach. A few time-oriented studies are also available for other response processes. For orcharding, see Faris (1960b); for multi-harvest crops, Eidman, Lingle and Carter (1963)—whose work forms the basis of Section 3.9.1 above; for beef feedlots, Faris (1960a) and Kennedy (1972 and 1973); for pigs, Battese et al. (1968), Dent (1964), Fawcett (1973) and Townsley (1968); for forests, Chisholm (1966 and 1975), Gaffney (1960) and Sinden (1964–5) who present an adaptation of the Faustmann criterion to answer the forest manager's problem of deciding whether to fell now or in five years or so. For irrigation, see Beringer (1961), Dorfman (1963), Hall and Buras (1961), Moore (1961), Ram (1963), Yaron (1971) and Yaron et al. (1963 and 1973); and the studies by Flinn and Musgrave (1967), Dudley, Howell and Musgrave (1971) and Minhas, Parikh and Srinivasan (1974) which are of special interest for their recognition of the problem of sequential input injections and the use of simulation procedures. Analyses partially covering time effects in milk response have been made by Dean (1960), Heady et al. (1964 a and b), O.E.C.D. (1965 and 1969a) and Paris et al. (1970). The economic appraisal of alternative crop rotations is considered by Battese and Fuller (1972), Battese, Fuller and Shrader (1972), Fuller (1965), Fuller and Cady (1965), and Shrader, Fuller and Cady (1966). They have also given consideration to the problem of input carry-over effects between response periods. More specifically time-oriented analysis of the crop cycling and fertilizer carry-over problem using dynamic programming is presented by Kennedy et al. (1973)—whose work we have followed in Section 3.9.4 above—and Stauber, Burt and Linse (1975). The actual measurement of residual or carry-over fertilizer has been discussed by, among others, Barrow (1973), Barrow and Campbell (1972), Colwell (1967–8), Stauber and Burt (1973), and Waugh, Cate and Nelson (1973). Extending the fertilizer carry-over question, Helyar and Godden (1976) have shown that soil nutrient status may be regarded as a capital asset whose size from year to year is significantly influenced by crop output and fertilizer input. This allows appraisal of optimal fertilizer use in the context of an allocation between annual variable and capital costs (Godden and Helyar, 1976).

As well as failing to allow properly for time effects, livestock-feed analyses based on pen-feeding trials have also generally involved *ad libitum* feeding. On *a priori* grounds based on considerations of diminishing returns, there are strong reasons for believing sub-*ad libitum* feeding levels must be optimal since the marginal product of feed is zero beyond stomach capacity and greater than zero below it. Empirical evidence of the non-optimality of *ad libitum* feeding is also available from grazing-stocking rate trials as analysed by Chisholm (1965) and Lloyd (1966), the optimal stocking rate being well beyond a grazing pressure which would allow an *ad libitum* feed level per animal. For pen-feeding trials and lot-feeding systems of production, Duloy and Battese (1967) have developed a time-oriented model based on feed input factors as a proportion of bodyweight. This model allows analysis both in terms of time opportunity costs and sub-*ad libitum* levels of feeding.

No adequate empirical studies of the pasture livestock–response complex are as yet available although a start has been made with the studies of beef response to pasture and soilage by Nelson and Castle (1958), Jones and Hocknell (1962), McConnen *et al.* (1963), McConnen (1965) and Heady *et al.* (1963a and b) and the pasture grazing stocking-rate studies of Bennett *et al.* (1970), Cannon (1969 and 1972), Conniffe *et al.* (1970), Harlan (1958), Hart (1972), Jones and Sandland (1974), Mauldon (1968) and Sandland and Jones (1975). These studies variously illustrate the underlying biological considerations and the alternative empirical approaches (i) of relating livestock product directly to fertilizer and the input of livestock (stocking rate), or (ii) of using a multi-stage (intermediate product) approach of superimposing a pasture–livestock production function on a fertilizer–pasture production function (perhaps also using linear or dynamic programming to impute a price to pasture and to allow for seasonal variations in pasture supply and demand). Some discussion of these alternatives of a direct or of an intermediate product approach has been given by Anderson (1967a) and Dillon and Burley (1961).

The rather elaborate and more realistic but somewhat hypothetical pasture–livestock grazing model of Section 3.9.3 is based on the work of Dillon and Burley (1961). Using dynamic programming, Throsby (1964b) has explored the problem of ascertaining the optimal time path of development for a pasture–grazing response process. Arcus (1963) and more recently Anderson (1974c), Dent and Anderson (1971), Greig (1972), Paltridge (1970), Reid and Thomas (1973) and Wright and Dent (1969) have discussed or investigated the pasture–livestock grazing problem using a systems-modelling simulation approach.

Like the analysis presented in the current chapter, virtually all the studies referred to above assume prices and yields are known with certainty. This is a simplification since crop and livestock response processes are inevitably risky. In consequence, they involve questions of subjective judgement and preferences about the risks involved. These influences on best operating conditions are outlined in Chapter 4.

3.11 Exercises

3.11.1. Derive best-operating conditions for the response process of equations (3.9) and (3.10) if it is to be run continuously under the constraint that output in each response period must equal some fixed amount Y^*.

3.11.2. Hoepner and Freund (1964, ch. 4) have estimated functions of the following form corresponding to the broiler process of equations (3.37) and (3.38):

$$Y = b_0 + b_1 C^2 + b_2 P + b_3 P^3 + b_4 t^2 + b_5 t^3 + b_6 CP^2 + b_7 C^2 P$$
$$+ b_8 C^2 t + b_9 P^2 t,$$

$$F = a_0 + a_1 C + a_2 P + a_3 P^2 + a_4 P^3 + a_5 t + a_6 t^2 + a_7 t^3$$
$$+ a_8 CP + a_9 CP^2 + a_{10} Ct + a_{11} C^2 t.$$

Using these functions, derive the criteria for best-operating conditions corresponding to equations (3.43).

3.11.3. Contrast the treatment of time effects as developed in this chapter with that used in Heady and Dillon (1961, chs. 8–13) and Heady *et al.* (1963 a and b).

3.11.4. Assess the relative amenability of polynomial, Mitscherlich, resistance, and power-type response functions for time-oriented analyses of the type developed in this chapter.

3.11.5. How adequate is the role assigned to time in the pasture-livestock model of Section 3.9.3?

3.11.6. Construct a table analogous to Table 3.1 but couched in terms of the criteria of equations (3.15), (3.24) and the Faustmann criterion of equation (3.23).

3.11.7. Extend the analysis of Section 2.8 for multiple response with input control so as to accommodate time opportunity cost and time preference effects.

3.11.8. Show the equivalence of equations (3.19) and (3.22).

3.11.9. Given the grain response function

$$Y = 31\cdot4 + 2\cdot9N - 0\cdot4N^2$$

with $V_t = 0\cdot2$ for t from 1 to 4, $R_4 = 1$, $\alpha = 0\cdot9$, $p_y = 10$ and $p_{Nt} = 2 + t$, use the dynamic programming procedure of Section 3.9.4 to determine the optimal fertilizer policy $\{N^*_i - R_i; i = 1, 2, 3, 4\}$ and optimal return $f_4(R_4 = 1)$ for a planning sequence of four crop years.

3.11.10. Extend the analysis of Section 3.9.4 to the case of two fertilizers each with carry-over between periods.

CHAPTER 4

Response Efficiency Under Risk

4.1 Introduction

Just as the influence of time complicates the analysis of response efficiency, so also does risk. Indeed, given the yield and price uncertainties that typically prevail in crop and livestock production, risk effects are generally more significant than time effects. They are also more difficult to analyse. Their difficulty of appraisal arises not so much in the mathematical sense (though that can be complicated enough) but because risk assessment is a personal matter. It involves personal or subjective judgement both about the chances to be associated with the different possible outcomes that might arise from any particular choice, and about preference between the sets of possible outcomes that are associated with alternative choices. Because of these elements of subjective judgement, the best operating conditions that would be appropriate for one decision maker may be quite inappropriate for another.

In the main we will consider risk effects in a quasi-timeless context so as to avoid complicating the analysis with the time-dependent considerations of Chapter 3. Note, however, that it is impossible to avoid an implicit time dimension in risk analysis. Present uncertainty about future outcomes necessarily implies a present and a future. Only with the passage of time between initiation of the response process and its termination is uncertainty resolved in terms of the choices made.

We will use the words RISK and UNCERTAINTY interchangeably to describe a decision situation which does not have a single sure outcome. Risky or uncertain choice in the response context thus implies a decision about input variables where the outcome in terms of profit is not certain because of uncertainty in yield or price or both. Further, as argued in

Section 4.3.2 below, we believe and adopt the view that risky or uncertain outcomes can always be described by a probability distribution and that this distribution is necessarily subjective. Risky choice thus implies judgements about and choices between alternative subjective probability distributions of outcomes.

As noted in the further reading to this chapter, a number of definitions of risk and a variety of approaches to handling it have been advocated. Chief among these suggestions have been such different procedures as: using various non-probabilistic criteria for choice derived from game theory; applying a risk discount factor to possible returns so as to ensure conservative analysis; using the expected value of returns for evaluation; working with a safety-first rule of requiring some minimum level of profit at a guaranteed level of probability; and using a decision theory approach based on the theorem of expected utility.

All of these approaches may have some relevance for different decision makers from a descriptive or behavioural point of view. In normative terms, however, they are not equally attractive. The game-theoretic procedures are wrong in ignoring the fact that subjective probabilities can always be ascribed to uncertain events. Though they have behavioural appeal to some decision makers, risk-discount factors and safety-first levels are arbitrary and have no logical foundation. The use of expected profit values to choose between alternatives implies indifference to the possible variation in returns.

Only the decision theory approach based on the maximization of expected utility, as outlined in Section 4.3.3 below, is normatively coherent and logical as a basis for risky choice. Accordingly, we will appraise risky response in terms of expected utility. This implies: first, the use of subjective probability distributions to specify the profit arrays associated with alternative choices about input use; and, second, the use of an objective function expressed not in terms of profit *per se* but in terms of a subjectively derived measure known as expected utility. After first outlining the sources of risk, we will sketch these concepts of subjective probability and expected utility, and then discuss their theoretical implications and use in the determination of best operating conditions for risky response.

4.2 Sources of Risk

Risky returns occur in the context of crop and livestock response processes because either yields or prices or both are uncertain.

4.2.1. YIELD UNCERTAINTY

Uncertainty about yield arises because invariably some input variables are not under the decision maker's control and their levels are not known at the time decisions have to be made about the controlled input variables. Consider a "timeless" response function which includes all the non-fixed input variables $X_1, X_2, \ldots, X_m$ which influence yield. From a decision point of view, these input factors may be classified into three groups as follows:

(a) $X_1, X_2, \ldots, X_n$ denoting variables whose levels are controlled by the decision maker; these are the DECISION VARIABLES.

(b) $X_{n+1}, X_{n+2}, \ldots, X_k$ denoting variables whose levels are not controlled but are known to the decision maker at the time he has to decide on the decision variables; these are the PREDETERMINED VARIABLES.

(c) $X_{k+1}, X_{k+2}, \ldots, X_m$ denoting variables whose levels are neither controlled by nor known to the decision maker at the time he chooses levels for the controlled inputs; these are the UNCERTAIN VARIABLES.

We can thus write the complete response function as

$$Y = f(X_1, \ldots, X_n; X_{n+1}, \ldots, X_k; X_{k+1}, \ldots, X_m). \quad (4.1)$$

Typical controlled variables in crop production are fertilizers, seed quantity, crop variety, herbicides and insecticides; and in livestock production, feed type and quantity, animal density and feed additives. Uncontrolled but predetermined input factors could be, for example, such site characteristics as initial soil fertility and soil moisture content in crop production and such animal characteristics as genetic merit in livestock. For both crop and livestock processes, the major uncertain (i.e. uncontrolled and unknown) inputs are such climatic variables as rainfall, temperature, wind, solar radiation, etc.

RESPONSE EFFICIENCY UNDER RISK   105

Yield uncertainty arises from the influence of the uncontrolled variables $X_{k+1}, \ldots, X_m$ whose levels are unknown. As equation (4.1) indicates, for each possible set of values of the uncertain inputs in conjunction with chosen values for the decision variables and given values for the predetermined variables, there will be some corresponding level of output. Since we do not know the input values $X_{k+1}, \ldots, X_m$, we cannot be sure of the yield we will obtain. As discussed later, however, we can always specify a subjective probability distribution for yield in relation to possible combinations of levels of the uncertain input variables. This probability distribution of yield will be relevant to the appraisal of best operating conditions so long as there is any interaction in response between any of the decision variables and the uncertain variables. If these two groups of variables have independent (i.e. only additive and not multiplicative) effects on response, the marginal products of the decision variables $X_1, \ldots, X_n$ and their best operating levels will be independent of the uncertain variables. This will not be the case if the decision variables interact with the uncertain variables.

Stated another way, if the probability distribution of yield associated with the uncertain variables $X_{k+1}, \ldots, X_m$ is independent of or not conditioned by the level of the decision variables $X_1, \ldots, X_n$, choice of levels for the decision variables can be made without regard to yield risk; no matter what levels are chosen for the decision variables, they will have no influence on the distribution of yield. Hence, best operating conditions for a particular process (though not the choice between alternative processes) can in this case be chosen without regard to yield uncertainty. Conversely, if the probability distribution of yield relative to $X_{k+1}, \ldots, X_m$ can only be specified conditional on $X_1, \ldots, X_n$, choice of $X_1, \ldots, X_n$ will influence the distribution of yield. Therefore, in this case, choice of best operating conditions should allow for yield risk effects.

In fact, crop and livestock response invariably involves interaction between the decision variables (such as fertilizer and feed) and the uncertain input variables (such as rainfall and temperature) whose levels are not known *a priori*. Accordingly, unless yield uncertainty is eliminated by, e.g. government-sponsored yield insurance schemes, risk effects associated with uncertain response will usually be pertinent to the appraisal of best operating conditions.

4.2.2. PRICE UNCERTAINTY

With allowance for fixed costs F, and ignoring time effects, net return or profit from the response process of equation (4.1) is given by

$$\pi = p_y Y - \sum p_i X_i - F \quad (i = 1, 2, \ldots, n) \quad (4.2)$$

where the input prices p_i are positive for the decision variables $i = 1, \ldots, n$ and zero for the uncontrolled variables $i = n + 1, \ldots, m$. Variable costs thus relate only to the decision variables. As equation (4.2) indicates, price uncertainty may be influential either through uncertainty about the product price p_y or uncertainty about the input prices p_i. In fact, by the nature of crop and livestock production, the product price p_y but not the input prices p_i will generally be uncertain at the time a decision has to be made about the level of the controlled input variables (though there may sometimes be price policy schemes which reduce or eliminate product price uncertainty). For our analysis, we will assume price uncertainty only occurs in relation to product price p_y. The input prices p_i will be taken as known constants.

4.3 Risk and the Objective Function

The importance of yield and price uncertainty lies in their influence on profit possibilities. At the same time, because of the great qualitative difference between situations of certainty and risk, they force a change in the form of the objective function. The simple riskless objective function of equation (2.2) is inadequate for risky response appraisal. What sort of objective function is required will become obvious from a consideration of profit uncertainty.

4.3.1. PROFIT UNCERTAINTY

Given our assumptions of known constant input prices, the profit equation (4.2) shows that profit uncertainty may arise because of uncertainty residing in either the product price p_y, the yield Y, the input quantities X_i or the fixed costs F. Inevitably there will be some slight uncertainty about the controlled input quantities X_i and fixed costs F. These effects, however can usually be safely ignored. We therefore assume that all uncertainty about profit π arises from yield and product price risks. Since product price and yield occur together in the profit equation (4.2) as the gross revenue term $p_y Y$, this can be very convenient for statistical appraisal—it means that the probability distribution of profit bears a direct relation to the distribution of gross revenue or the joint probability distribution of p_y and Y.

Though the probability distribution of p_y is unlikely to be influenced by the decision variables $X_1, \ldots, X_n$, this is not true for the distribution of X—as was discussed in Section 4.2.1 above. Accordingly, the joint distribution of p_y and Y, and hence the distribution of π, will be conditional on the decision variables $X_1, \ldots, X_n$. Thus we need to write the probability distribution of profit, denoted $h(\pi)$, in the conditional form

$$h(\pi | X_1, \ldots, X_n) = h(p_y Y - \sum p_i X_i - F | X_1, \ldots, X_n). \quad (4.3)$$

Since $\sum p_i X_i$ and F are constant for any given set of $X_1, \ldots, X_n$ values, the distribution of $h(\pi | X_1, \ldots, X_n)$ will be of the same shape as the distribution $g(p_y Y | X_1, \ldots, X_n)$ of gross revenue. The only difference between the two distributions is that the mean of the profit distribution will be lower by the amount $(\sum p_i X_i + F)$ than the mean of the gross revenue distribution.

Choice of best operating conditions for the decision variables $X_1, \ldots, X_n$ therefore devolves to a choice between alternative probability distributions of profit (or gross revenue). What we need in order to make such choices is some criterion which enables us to rank alternative probability distributions of profit. Such a criterion, which has the attractive feature of being normatively logical for many decision makers, is given by the expected utility theorem.

4.3.2. EXPECTED UTILITY AND SUBJECTIVE PROBABILITY

The expected utility theorem is based on three axioms or postulates which many people regard as reasonable bases of choice. These three axioms are:

Ordering. A person either prefers one of two probability distributions h_1 or h_2, or is indifferent between them. Further, if a person prefers

h_1 to h_2 (or is indifferent between them) and prefers h_2 to h_3 (or is indifferent between them), then he will prefer h_1 to h_3 (or be indifferent between them).

Continuity. If a person prefers the probability distribution h_1 to h_2 to h_3, then there exists a unique probability p such that he is indifferent between h_2 and a lottery with a probability p of yielding the distribution h_1 and a probability $(1-p)$ of yielding the distribution h_3.

Independence. If the distribution h_1 is preferred to h_2, and h_3 is some other probability distribution, then a lottery with h_1 and h_3 as prizes will be preferred to a lottery with h_2 and h_3 as prizes if the probability of h_1 and h_2 occurring is the same in both cases.

As can be shown, these three axioms imply the EXPECTED UTILITY THEOREM which states that: for a decision maker whose preferences do not violate the axioms of ordering, continuity and independence, there exists both (a) a unique subjective probability distribution for the set of outcomes associated with any risky choice alternative that he faces; and (b) a function U, called a UTILITY FUNCTION, which gives a single-valued utility index or measure of attractiveness for each of the risky alternatives that he faces.

The implied subjective probabilities follow the usual laws of probability. The utility function U has the following properties:

(i) If the probability distribution h_1 is preferred to h_2, then the utility index of h_1 will be greater than the utility index of h_2, i.e. $U(h_1) > U(h_2)$. Conversely $U(h_1) > U(h_2)$ implies h_1 is preferred to h_2.

(ii) If A is an act or choice with a set of uncertain outcomes $\{a\}$, then the utility of A is equal to the statistically expected utility of A where the expectation is taken in terms of the subjective probability distribution $h(a)$ implied by the expected utility theorem. Thus, using E to denote mathematical expectation,

$$U(A) = E[U(A)] \qquad (4.4)$$

$$= \int_{-\infty}^{\infty} U(a)h(a)d(a). \qquad (4.5)$$

As these equations indicate, only the mean or expected value of utility is relevant for choice; the expected value of utility takes full account of all the attributes (mean, variance, skewness, etc.) of the probability distribution $h(a)$ of outcomes.

(iii) Uniqueness of the function U is only defined up to a positive linear transformation. Given a function U, any other function U^t such that

$$U^t = \alpha_1 U + \alpha_2, \quad \alpha_1 > 0, \quad (4.6)$$

will serve as well as the original function. Thus utility is measured on an arbitrary scale and is a relative measure analogous, for example, to the various scales used for measuring temperature. Further, because there is no absolute scale for utility and because a decision maker's utility function reflects his own personal valuations, it is impossible to compare one person's utility indices with another's.

Note the remarkable nature of the expected utility theorem. On the basis of three simple and reasonable postulates about rational choice, it implies: (a) the existence of a subjective probability distribution for the uncertain outcomes associated with any risky alternative a decision maker may be contemplating; (b) a utility function that reflects the decision maker's preferences between alternative risky choices; and (c) that risky choice is optimized by choosing the alternative with the highest expected utility index. The utility approach thus brings together in an explicit way the two crucial elements in risky choice—the decision maker's personal degrees of belief and his personal degrees of preference. For these reasons, practical difficulties aside, expected utility is a most attractive measure for appraising risky response alternatives in crop and livestock production.

4.3.3. UTILITY OBJECTIVE FUNCTION

Taking the maximization of expected utility as our criterion for risky choice, the objective function must be formulated in utility terms. Thus we wish to choose levels for the decision variables $X_1, \ldots, X_n$ of the response function (4.1) so as to maximize the utility objective function specified by:

$$U = U(\pi) \tag{4.7}$$
$$= E[U(\pi)] \tag{4.8}$$
$$= \int_{-\infty}^{\infty} U(\pi) h(\pi | X_1, \ldots, X_n) d(\pi) \tag{4.9}$$

where equation (4.8) follows from equation (4.7) by virtue of the expected utility theorem, profit π is as defined by equation (4.2) but on a total enterprise—not a technical unit—basis, and $h(\pi|X_1, \ldots, X_n)$ is the subjective probability distribution of profit (conditional on the level of the decision variables) implied by the expected utility theorem. Since a particular profit outcome, say π', implies a corresponding particular utility outcome $U(\pi')$, the probability distribution of profit $h(\pi|X_1, \ldots, X_n)$ is also the probability distribution of possible utility values with, of course, the scale of the random variable π transformed to that of U via the utility function.

Some general aspects of the utility function for profit, $U(\pi)$, are as follows. First, $U(\pi)$ may have any algebraic form so long as it is monotonically increasing over the range of interest. Equivalently, it should have $dU/d\pi > 0$ reflecting a positive marginal utility for profit. This corresponds to more profit being preferred to less. Commonly used forms of the utility function which satisfy this requirement are the quadratic

$$U = \pi + b\pi^2 \tag{4.10}$$

which requires $\pi > -1/2b$ if $b > 0$ and $\pi < -1/2b$ if $b < 0$; the logarithmic function

$$U = \log_e(W + \pi) \tag{4.11}$$

where W is the decision maker's current wealth; and the power function

$$U = (W + \pi)^c, \quad c > 0. \tag{4.12}$$

Second, the utility function at a given level of π indicates risk aversion, risk indifference or risk preference according as $d^2U/d\pi^2$ is less than, equal to, or greater than zero, respectively. Thus the quadratic utility function (4.10) exhibits risk aversion if $b < 0$, risk indifference if $b = 0$, and risk preference if $b > 0$. As a point of empirical fact, most decision makers are risk averse.

Third, unless the utility function is linear (i.e. $d^2U/d\pi^2 = 0$), it follows that $U(k\pi) \neq kU(\pi)$ for any non-zero constant $k \neq 1$, and also that

$U(\pi_1 + \pi_2) \neq U(\pi_1) + U(\pi_2)$ for any non-zero profits π_1 and π_2. In consequence, the utility appraisal of risky response must be based not on the gross margin $(p_y Y - \sum p_i X_i)$ but on profit or net return $(p_y Y - \sum p_i X_i - F)$, i.e. fixed costs must be subtracted out. Likewise, utility must be assessed on a total profit (i.e. total enterprise) basis and not on a technical unit basis (i.e. per hectare or per animal). Thus, if we are dealing with a crop or livestock response process involving H hectares or animals and profit as per equation (4.2) is expressed on a per hectare or per animal basis, we must convert equation (4.2) to a total profit basis by multiplying it by the factor H. To avoid this complication, we will—unless otherwise noted—assume profit as per equation (4.2) is already expressed on a total basis so that the utility objective function of equation (4.9) is on a total profit basis. The above requirements contrast with the riskless analysis of Chapter 2 in which fixed costs can be ignored and efficiency assessed on a technical unit basis.

Fourth, just as interpersonal comparisons of utility are impossible, it must also be emphasized that a decision maker's utility function *per se* cannot be the subject of normative assessment. The utility function simply describes his preferences; it cannot be classified as good or bad, right or wrong, efficient or inefficient.

Fifth, by taking the expectation of a Taylor series expansion of the utility function, the (expected) utility objective function (4.9) may be expressed as a function of the moments of the profit distribution. Thus

$$U = f(E(\pi), V(\pi), S(\pi), \ldots) \quad (4.13)$$

where $E(\pi)$, $V(\pi)$, $S(\pi)$, ..., respectively, denote the mean, variance, skewness and higher moments $M_k(\pi) = E[\pi - E(\pi)]^k$ about the mean of profit. As an empirical matter, in general the higher the moment, the less important its influence in utility appraisal. Rarely is it necessary to consider moments beyond the third, i.e. effects corresponding to skewness; and for many decision makers and many decision problems, reasonable or adequate appraisal is given by consideration of just the mean and variance of profit.

Figures 4.1(a) and (b) illustrate the utility concept with quadratic utility functions for profit measured in thousands of dollars. These functions respectively exhibit (a) risk aversion (the most common case) and (b) risk preference.

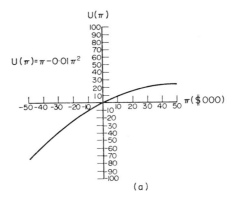

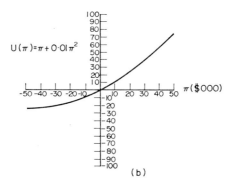

Fig. 4.1. Example of utility functions exhibiting (a) risk aversion and (b) risk preference.

4.4 Best Operating Conditions under Risk

With profit specified on a total enterprise basis as per equation (4.2) and the utility objective function specified as per equation (4.9), we are now in a position to at least theoretically assess best operating conditions under risk. For simplicity of exposition we begin with the case of response involving only a single input decision variable.

4.4.1. SINGLE DECISION VARIABLE

Suppose we wish to determine best operating conditions for the response process

$$Y = f(X_1; X_2, \ldots, X_k; X_{k+1}, \ldots, X_m) \quad (4.14)$$

with a single decision variable X_1; predetermined variables $X_2, \ldots, X_k$; and uncertain variables $X_{k+1}, \ldots, X_m$ which interact with X_1 so that the probability distribution of yield is conditional on X_1. No uncertainty attaches to fixed costs F or to the level of the decision variable X_1 or its unit price p_1. Product price per unit p_y is uncertain. Reflecting the uncertainty of both p_y and Y, total profit as defined by

$$\pi = p_y Y - p_1 X_1 - F \quad (4.15)$$

is a random variable with some subjective probability distribution $h(\pi | X_1)$.

The utility objective function to be maximized may be expressed as

$$U = \int_{-\infty}^{\infty} U(\pi) h(\pi | X_1) d(\pi). \quad (4.16)$$

However, it is more convenient for analysis to express utility in the moment form of equation (4.13). To keep the analysis simple, we make the not-unreasonable assumption that the mean and variance of profit are the only relevant probability parameters for utility appraisal. Thus we have

$$U = f(E(\pi), V(\pi)). \quad (4.17)$$

The first-order condition for maximizing U is that the derivative

$$dU/dX_1 = [\partial U/\partial E(\pi)] [dE(\pi)/dX_1] + [\partial U/\partial V(\pi)] [dV(\pi)/dX_1] \quad (4.18)$$

be equal to zero. This implies

$$0 = dE(\pi)/dX_1 + \{[\partial U/\partial V(\pi)]/[\partial U/\partial E(\pi)]\}[dV(\pi)/dX_1], \quad (4.19)$$

where the term in curly brackets is the negative of the RATE OF SUBSTITUTION IN UTILITY of $E(\pi)$ for $V(\pi)$, written RSU_{EV} and defined as the slope $dE(\pi)/dV(\pi)$ of an isoutility curve in mean-variance profit space. That the curly-bracket term is the negative of RSU_{EV} can be seen by rearrangement of the total implicit differential

$$dU = [\partial U/\partial E(\pi)][dE(\pi)/dV(\pi)] + \partial U/\partial V(\pi) \quad (4.20)$$

of $U = f(E(\pi), V(\pi))$ with U fixed at U^* so that $dU = 0$. Hence

$$RSU_{EV} = [dE(\pi)/dV(\pi)]_{U=U^*} \quad (4.21)$$
$$= -[\partial U/\partial V(\pi)]/[\partial U/\partial E(\pi)]_{U=U^*}. \quad (4.22)$$

Thus the utility maximizing criterion of equation (4.19) may be rearranged as

$$RSU_{EV} = [dE(\pi)/dX_1]/[dV(\pi)/dX_1] \quad (4.23)$$
$$= dE(\pi)/dV(\pi) \quad (4.24)$$

which says that, when mean and variance of profit are the only relevant probability parameters, best operating conditions under risk imply equality between (a) the rate of substitution in utility of $E(\pi)$ for $V(\pi)$, and (b) the rate of substitution in response of $E(\pi)$ for $V(\pi)$. Alternatively stated, we require tangency between (a) an isoutility curve in mean-variance profit space, and (b) the mean-variance frontier of response possibilities depicted in mean-variance profit space. Figure 4.2 illustrates the criterion, best operating conditions being achieved with the level of X_1 corresponding to the tangency point A on the mean-variance frontier of response possibilities OAB. At this point, maximum utility of U_2^* is obtained. Note that response possibilities involving combinations of $E(\pi)$ and $V(\pi)$ lying above OAB are not feasible.

To proceed further with the utility maximizing criterion of equations (4.19) or (4.24), we need to express the derivatives of $E(\pi)$ and $V(\pi)$ with respect to X_1 in terms of our basic economic variables. To do this we take the mean and variance of π as implied by equation (4.15). Thus we have

$$E(\pi) = E(p_y)E(Y) - p_1 X_1 - F \quad (4.25)$$
$$V(\pi) = [E(p_y)]^2 V(Y) + [E(Y)]^2 V(p_y) + V(p_y)V(Y), \quad (4.26)$$

where, in equation (4.26), we have made the reasonable assumption that product price p_y and yield Y are statistically independent relative to the individual decision maker. Taking the derivatives

$$dE(\pi)/dX_1 = E(p_y)dE(Y)/dX_1 - p_1 \quad (4.27)$$
$$dV(\pi)/dX_1 = \{[E(p_y)]^2 + V(p_y)\}[dV(Y)/dX_1]$$
$$+ 2V(p_y)E(Y)[dE(Y)/dX_1] \quad (4.28)$$

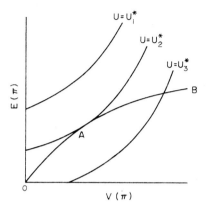

FIG. 4.2 Solution of equation (4.24) at tangency between a mean-variance isoutility curve (U_2^*) and the mean-variance frontier (OAB) of response possibilities.

from equations (4.25) and (4.26), respectively, and substituting them into equation (4.19), we have

$$E(p_y)[dE(Y)/dX_1] = p_1 + (RSU_{EV})[dV(\pi)/dX_1] \quad (4.29)$$

or, more fully,

$$E(p_y)[dE(Y)/dX_1] = p_1 + (RSU_{EV})[\{[E(p_y)]^2 + V(p_y)\} \\ [dV(Y)/dX_1] + 2V(p_y)E(Y)[dE(Y)/dX_1]]. \quad (4.30)$$

This expression reflects the effects of both yield and price uncertainty. For the risky response process we are considering, it constitutes the usual economic ground rule that efficiency necessitates equality between, however measured, marginal revenue and marginal costs. The LHS of equation (4.30) expresses marginal revenue as the expected value of marginal product per unit of X_1. Marginal cost on the RHS consists of the direct marginal cost per unit of X_1, i.e. p_1, plus a MARGINAL COST OF RISK per unit of X_1 due to profit variance. This marginal cost of risk consists of RSU_{EV} weighted by a factor involving the mean $E(p_y)$ and variance $V(p_y)$ of price, the mean yield $E(Y)$, the marginal variance of yield $dV(Y)/dX_1$, and the marginal expected product $dE(Y)/dX_1$. The decision maker's personal evaluation of risk is reflected in equation (4.29) both by RSU_{EV} (which is determined by his utility function), and by the mean and variance parameters of his subjective probability distributions for p_y and Y.

Intuitively, it is obvious that for a risk-averse decision maker, variance constitutes a friction interpretable as equivalent to a risk-induced increase in input cost. Hence, under risk aversion, risk implies a lower level of input use than in the absence of risk. With risk preference, the reverse would apply. And with risk indifference, best operating conditions would be as in the riskless case. That these intuitions are correct can be seen from consideration of the expression for the marginal cost of risk in equation (4.30). The RSU_{EV} term will be positive, zero or negative according as the decision maker's utility function exhibits risk aversion, indifference or preference, respectively. With risk aversion, for example, RSU_{EV} is positive since any increase in variance must be compensated for by an increase in expected profit if utility is to remain unchanged. Thus for the quadratic expected utility function

$$U = E(\pi) + bE(\pi^2) \tag{4.31}$$

$$= E(\pi) + b[E(\pi)]^2 + bV(\pi) \tag{4.32}$$

which corresponds to equation (4.10), we have

$$RSU_{EV} = -b/[1 + 2bE(\pi)] \tag{4.33}$$

which is positive, zero or negative according as the utility function exhibits risk aversion $(b < 0)$, indifference $(b = 0)$, or preference $(b > 0)$ within its relevant range. The other more complicated term comprising the marginal cost of risk in equation (4.30) will be positive so long as yield variance increases with input use, i.e. $dV(Y)/dX_1 > 0$, and marginal expected product $dE(Y)/dX_1$ is positive. Both these conditions will normally prevail within the relevant range of operation of the response process. Sometimes it may be that $dV(Y)/dX_1$ is negative; for example, if X_1 is irrigation water or insecticide in a very drought or insect prone situation. But even in such cases, the negative effect of $dV(Y)/dX_1$ must outweigh the positive effect of the $2V(p_y)E(Y)[dE(Y)/dX_1]$ cost term of equation (4.30) before risk will induce enhanced input use. Should this occur, it is logical in that we would expect a risk-averse decision maker to use more of an input if in that way he can reduce yield risk.

With only yield risky, i.e. p_y a known constant, equation (4.30) collapses to the simpler form

$$p_y[dE(Y)/dX_1] = p_1 + (RSU_{EV})[p_y{}^2 dV(Y)/dX_1]. \quad (4.34)$$

The optimal level of X_1 is then that at which the value of its marginal expected product (LHS) is equal to its direct marginal cost p_1 plus a marginal cost of yield risk which consists of the yield-induced marginal variance of revenue weighted by RSU_{EV}.

Likewise if only product price is uncertain, i.e. Y a constant for given X_1 so that $V(Y)$ is zero, we have

$$E(p_y)[dY/dX_1] = p_1 + (RSU_{EV})2V(p_y)YdY/dX_1. \quad (4.35)$$

In this case best operating conditions imply equality between the expected value of marginal product (LHS) and marginal cost composed of the direct marginal factor cost p_1 plus a marginal cost of price variability.

Comparing the marginal costs of risk in equation (4.30) with the sum of those in equations (4.33) and (4.34), it may be noted that when both price and yield risk are present, the marginal cost of risk is increased by the price-yield risk-interaction element $(RSU_{EV})V(p_y)dV(Y)/dX_1$.

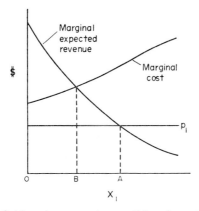

FIG. 4.3. Effect of risk on best operating conditions for a risk-averse decision maker as per equations (4.30), (4.34) or (4.35).

Figure 4.3 illustrates the general nature of the effect of risk on best operating conditions for a risk-averse decision maker as implied by equations (4.30), (4.34) or (4.35). The difference between the marginal cost curve and the p_1 line is the marginal cost of risk. In the absence of risk, the optimal quantity of X_1 would be OA; with risk it is OB.

So far we have made no mention of the second-order condition $d^2U/dX_1^2 < 0$ required for maximization of the utility objective function (4.9) or (4.17). Nor will we do so except in the further reading of Section 4.7. The conditions are complex and have no insightful implications. Empirically, the second-order conditions will tend to be automatically taken into account, as will input constraints, by the necessity to use numerical methods of analysis.

Example of Single-factor Risk Analysis

Suppose a farmer plans to grow 100 hectares of maize. His utility function for profit is

$$U = \pi - 0 \cdot 00002\pi^2. \tag{4.36}$$

Fixed costs are $100 per hectare. His subjective probability distribution for the price of maize per kilogram has a mean of $0·04 and a variance of 0·0001. The decision variable of interest is nitrogen fertilizer priced at $0·30 per kilogram. Based on historical data, he judges that the mean and variance of yield per hectare can be expressed as

$$E(Y/ha) = 6000 + 30N - 0 \cdot 1N^2, \tag{4.37}$$

$$V(Y/ha) = 800000 + 10000N, \tag{4.38}$$

where N is kilograms of nitrogen per hectare. These two equations reflect the fact that the probability distribution of yield arising from uncertain climatic and other factors is conditional on N. Converting $E(Y)$ and $V(Y)$ to a total enterprise basis of 100 hectares, as required for the utility objective function, we have

$$E(Y) = 6(10^5) + 3(10^3)N - 10N^2 \tag{4.39}$$

$$V(Y) = 8(10^9) + (10^8)N. \tag{4.40}$$

Corresponding to equations (4.25) and (4.26) we thus have, on a 100 hectare basis,

$$E(\pi) = (0{\cdot}04)[6(10^5) + 3(10^3)\mathcal{N} - 10\mathcal{N}^2] - 30\mathcal{N} - 10^4 \quad (4.41)$$

$$V(\pi) = (0{\cdot}04^2)[8(10^9) + (10^8)\mathcal{N}] + [6(10^5) + 3(10^3)\mathcal{N} - 10\mathcal{N}^2]^2$$
$$(0{\cdot}0001) + (0{\cdot}0001)[8(10^9) + (10^8)\mathcal{N}]. \quad (4.42)$$

To determine best operating conditions we could proceed directly to solve the "marginal revenue equals marginal cost" criterion of either equation (4.23) or (4.29) using the relations of equations (4.33), (4.36), (4.41) and (4.42). It is not necessary to elaborate the utility objective function (4.32). For completeness, however, we may note that this utility objective function would be specified as

$$U = 0{\cdot}04[6(10^5) + 3(10^3)\mathcal{N} - 10\mathcal{N}^2] - 30\mathcal{N} - 10^4$$
$$- 0{\cdot}00002\{0{\cdot}04[6(10^5) + 3(10^3)\mathcal{N} - 10\mathcal{N}^2] - 30\mathcal{N} - 10^4\}^2$$
$$- 0{\cdot}00002\{0{\cdot}04^2[8(10^9) + (10^8)\mathcal{N}] + [6(10^5) + 3(10^3)\mathcal{N}$$
$$- 10\mathcal{N}^2]^2 0{\cdot}0001 + 0{\cdot}0001[8(10^9) + (10^8)\mathcal{N}]\}. \quad (4.43)$$

Given this equation, we could proceed to set $dU/d\mathcal{N}$ equal to zero and solve for the optimal level of $\mathcal{N}$. However, we will follow our development of equation (4.30).

Taking the necessary derivatives from equations (4.39) and (4.40), and making the other necessary substitutions, corresponding to the criterion of equation (4.30) we have:

$$0{\cdot}04[3(10^3) - 20\mathcal{N}] = 30 + 0{\cdot}00002[1 - 0{\cdot}00004\{0{\cdot}04$$
$$[6(10^5) + 3(10^3)\mathcal{N} - 10\mathcal{N}^2] - 30\mathcal{N} - 10^4\}]^{-1}\{(0{\cdot}04^2 +$$
$$0{\cdot}0001)10^8 + 2(0{\cdot}0001)[6(10^5) + 3(10^3)\mathcal{N} - 10\mathcal{N}^2][3(10^3)$$
$$- 20\mathcal{N}]\}. \quad (4.44)$$

Solving this equation graphically, as sketched in Fig. 4.4, indicates an optimal $\mathcal{N}$ of 74 kilograms per hectare. Without consideration of risk, the optimal $\mathcal{N}$ would be 112·5 kilograms per hectare. If only yield were uncertain, the optimal level of $\mathcal{N}$ would be 96. With only price uncertainty, it would be 93.

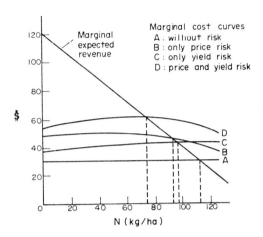

Fig. 4.4. Graphical solution of equations (4.34), (4.35) and (4.30) respectively illustrating the empirical example with only yield risk, with only price risk, and with both yield and price risk.

The marginal revenue and marginal cost for the 100 hectares of maize at various levels of N per hectare with only yield risk or only price risk or both are as shown in Table 4.1. These costs, when added to the direct marginal cost of N of $30 per 100 hectares, correspond respectively to the RHS of equations (4.34), (4.35) and (4.30) and the marginal cost curves C, B and D of Fig. 4.4. In this particular example, price risk is more significant than yield risk, and the interaction effect between price risk and yield risk is quite small, e.g. at $N = 80$ it is only $[30·98 - (12·80 + 17·38)] = \$0.80$ for the 100 hectares.

4.4.2. MULTIPLE DECISION VARIABLES

With multiple decision variables $X_1, \ldots, X_n$ and uncertain variables $X_{k+1}, \ldots, X_m$, we have the risky response function

$$Y = f(X_1, \ldots, X_n; X_{n+1}, \ldots, X_k; X_{k+1}, \ldots, X_m), \quad (4.45)$$

the risky profit function

$$\pi = p_y Y - \sum p_i X_i - F \quad (i = 1, 2, \ldots, n), \quad (4.46)$$

TABLE 4.1
MARGINAL REVENUE AND MARGINAL COST IN THE EMPIRICAL EXAMPLE OF SINGLE-FACTOR RISK ANALYSIS

Nitrogen per hectare	Marginal revenue	Direct marginal cost	Marginal cost of risk with:		
			Yield risk only	Price risk only	Yield and price risk
kg			$ on 100 ha		
0	120	30	7.26	16·34	24·06
40	88	30	10·00	19·36	29·98
80	56	30	12·80	17·38	30·98
120	24	30	13·33	8·16	22·32

and the utility objective function

$$U = = \int_{-\infty}^{\infty} U(p_y Y - \sum p_i X_i - F) h(\pi | X_1, \ldots, X_n) d(\pi). \quad (4.47)$$

As in the single-decision variable case, all these equations are formulated on a total enterprise and not a technical unit basis. Making the same assumption as in the single variable case that $E(\pi)$ and $V(\pi)$ are the only probability parameters of relevance, best operating conditions correspond to the set of X_i values which satisfy the set of n simultaneous equations

$$0 = \partial E(\pi)/\partial X_i - (RSU_{EV})[\partial V(\pi)/\partial X_i] \quad (4.48)$$

akin to equation (4.19). Alternatively, these first-order conditions for maximizing utility may be expressed analogously to equation (4.30) as the set of n "marginal revenue equals marginal cost" equations

$$E(p_y)[\partial E(Y)/\partial X_i] = p_i + (RSU_{EV})[\{[E(p_y)]^2 + V(p_y)\}]$$
$$[\partial V(Y)/\partial X_i] + 2V(p_y)E(Y)[\partial E(Y)/\partial X_i]]. \quad (4.49)$$

Each of these equations collapses to

$$p_y[\partial E(Y)/\partial X_i] = p_i + (RSU_{EV})[p_y^2 \partial V(Y)/\partial X_i] \quad (4.50)$$

if there is only yield risk; and to

$$E(p_y)[\partial Y/\partial X_i] = p_i + (RSU_{EV}) 2V(p_y) Y(\partial Y/\partial X_i) \quad (4.51)$$

if there is only price risk. As consideration of these equations (4.49), (4.50) and (4.51) indicates, the implications of risk in the multiple decision variable case are exactly analogous to the single variable case. In general, risk as measured by variance increases the marginal cost of input use for a risk-averse decision maker, thereby leading to overall lower input use than in the absence of risk. With risk preference, the reverse tends to occur. Since most decision makers are risk averse, the overall effect is of profit variability as a friction to input use.

4.4.3. CONSTRAINED MULTIPLE RESPONSE

As an example of constrained multiple response analysis under risk, suppose H hectares of homogeneous crop land are available for allocation between r different independent crop-response processes. Each crop exhibits yield risk but there is no price uncertainty. The risky decision problem is to choose how much land to give to each crop and, for each crop, what array of controlled input levels to use.

Utility must be assessed on the basis of the total area H. However, since we are concerned with the allocation of H between the alternative crops, it is most convenient in this particular type of problem to express the crop response functions on a technical unit basis. Thus, on a per hectare basis, we have the set of r response functions

$$Y_h = f_h(X_1, \ldots, X_n; X_{n+1}, \ldots, X_k; X_{k+1}, \ldots, X_m)$$
$$(h = 1, 2, \ldots, r) \qquad (4.52)$$

where X_1 to X_n are the input decision variables and X_{n+1} to X_m are the uncontrolled variables of which X_{k+1} to X_m are uncertain. We suppose the input decision variables are unconstrained and that, from relevant historical data, the mean yield per hectare $E(Y_h)$ and variance of yield per hectare $V(Y_h)$ are known for each crop as functions of the decision variables $X_1, \ldots, X_n$.

Total profit, subject to the area constraint H, is given by

$$\pi = \sum_h H_h(p_h Y_h - \sum_i p_i X_{ih}) - F, \qquad \sum H_h \leq H, \qquad (4.53)$$

where H_h is the area devoted to the hth crop; p_h is the unit price of Y_h; p_i is the unit price of the ith input decision variable X_i ($i = 1, 2, \ldots, n$);

X_{ih} is the amount of X_i used per hectare of the hth crop; and F is fixed cost on the H hectares.

The utility objective function is

$$U = \int_{-\infty}^{\infty} U(\pi)h(\pi|X_{11},\ldots,X_{nr})d(\pi) + \lambda(\sum H_h - H). \quad (4.54)$$

As this function indicates, the probability distribution of profit is conditional on the set of rn values X_{ih} that prevail. The utility function is also constrained to meet the restriction that land cannot exceed H hectares.

Again assuming that $E(\pi)$ and $V(\pi)$ are the only probability parameters of relevance, we can rewrite equation (4.54) as

$$U = f(E(\pi), V(\pi)) + \lambda(\sum H_h - H) \quad (4.55)$$

with

$$E(\pi) = \sum_h H_h[p_h E(Y_h) - \sum_i p_i X_{ih}] - F \quad (4.56)$$

and

$$V(\pi) = \sum_h H_h^2 p_h^2 V(Y_h) + 2 \sum_h \sum_{g>h} \rho_{hg} H_h H_g p_h p_g [V(Y_h)V(Y_g)]^{.5},$$
$$(h = 1, 2, \ldots, r; g = 2, 3, \ldots, r), \quad (4.57)$$

where ρ_{hg} is the correlation coefficient between the random variables Y_h and Y_g, i.e. between the yields of the hth and gth crops.

The decision quantities consist of the rn input levels X_{ih} and the r crop areas H_h, i.e. a total of $r(n+1)$. As usual, this set of best operating conditions (assuming satisfaction of the second-order conditions) is found by simultaneous solution of the set of first derivatives of the objective function with respect to the decision variables. From the utility objective function (4.55), for $\partial U/\partial X_{ih}$ we have the rn equations

$$\partial E(\pi)/\partial X_{ih} - (RSU_{EV})[\partial V(\pi)/\partial X_{ih}] = 0 \quad (4.58)$$

implying equality between RSU_{EV} and the rate of substitution in response of $E(\pi)$ for $V(\pi)$ for each input decision variable X_{ih}; for $\partial U/\partial H_h = 0$ we have the r equations

$$\partial E(\pi)/\partial H_h - (RSU_{EV})[\partial V(\pi)/\partial H_h] + \lambda = 0 \quad (4.59)$$

implying that if the land constraint is effective, RSU_{EV} exceeds the rate of substitution in response of $E(\pi)$ for $V(\pi)$ by the quantity $\lambda/[\partial V(\pi)/\partial H_h]$ for each crop area H_h; and for $\partial U/\partial \lambda = 0$, the single equation

$$\sum H_h - H = 0 \tag{4.60}$$

ensuring that the land constraint is met.

Eliminating λ from equations (4.59) and making the appropriate substitutions for the relevant derivatives of $E(\pi)$ and $V(\pi)$ from equations (4.56) and (4.57) respectively, we obtain the set of $r(n+1)$ equations:

$$p_h[\partial E(Y_h)/\partial X_{ih}] - p_i - (RSU_{EV})\{H_h p_h^2[\partial V(Y_h)/\partial X_{ih}]$$
$$+ \rho_{hg} H_g p_h p_g [V(Y_g)/V(Y_h)]^{.5}[\partial V(Y_h)/\partial X_{ih}]\} = 0 \tag{4.61a}$$

$$p_h E(Y_h) - \sum_i p_i X_{ih} - (RSU_{EV}) 2 p_h \{H_h p_h V(Y_h)$$
$$+ \rho_{hg} H_g p_g [V(Y_h) V(Y_g)]^{.5}\} = p_g E(Y_g) - \sum_i p_i X_{ig}$$
$$- (RSU_{EV}) 2 p_g \{H_g p_g V(Y_g) + \rho_{hg} H_h p_h [V(Y_h) V(Y_g)]^{.5}\} \tag{4.61b}$$

$$\sum H_h = H. \tag{4.61c}$$

These $r(n+1)$ equations are made up of rn in equations (4.61a) for the rn derivatives $\partial U/\partial X_{ih}$ from equations (4.58); $(r-1)$ in equations (4.61b) corresponding to $\partial U/\partial H_h = \partial U/\partial H_g$ for $g \neq h = 1, 2, \ldots, r$) from equations (4.59); and equation (4.61c) from equation (4.60). Obviously, simultaneous solution of these equations (even if there are no boundary solutions $X_{ih} = 0$ or $H_h = 0$) to find the set of rn values X_{ih} and the r values H_h for best operating conditions will be a messy affair. And much more so if account should also have to be taken of price uncertainty.

4.4.4. TIME AND RISK TOGETHER

To illustrate the implications of time and risk influences occurring together, we will extend the analysis of time-dependent response with time preference of Section 3.6.2 to incorporate price uncertainty. The time-dependent response process is specified by equations (3.9) and (3.10). Assuming mean and variance of profit provide an adequate representation of risk and that the mean $E(p_y)$ and variance $V(p_y)$ of product price are known subjectively, we have

$$E(\pi^{**}) = E(\pi)\rho/[(1+r)^t - 1] \tag{4.62}$$

$$= [E(p_y)\Upsilon - (\sum p_i X_i + F)(1+r)^t]\rho/[(1+r)^t - 1] \tag{4.63}$$

$$V(\pi^{**}) = V(\pi)\rho^2/[(1+r)^t - 1]^2 \tag{4.64}$$

$$= V(p_y)\Upsilon^2\rho^2/[(1+r)^t - 1]^2 \tag{4.65}$$

where the notation follows that of Section 3.6.2, i.e. π^{**} is the equivalent steady rate of profit flow per unit of time and ρ, equal to ln $(1+r)$, is the continuous compounding rate equivalent to r.

The utility objective function

$$U = f(E(\pi^{**}), V(\pi^{**})), \tag{4.66}$$

specified in terms of the flow of profit per unit of time equivalent to the lump-sum profit π, implies best operating conditions when we have either

$$\partial E(\pi^{**})/\partial X_i - (RSU_{EV})\partial V(\pi^{**})/\partial X_i = 0 \tag{4.67}$$

or, since $X_i = f_i(t)$,

$$\partial E(\pi^{**})/\partial t - (RSU_{EV})\partial V(\pi^{**})/\partial t = 0. \tag{4.68}$$

Making the appropriate substitutions into equation (4.68) of derivatives from equations (4.62) and (4.64) gives the condition

$$[\partial E(\pi)/\partial t]/(1+r)^t = E(\pi^{**}) + (RSU_{EV})\{\rho/[(1+r)^t - 1]\}$$
$$\{[\partial V(\pi)/\partial t](1+r)^{-t} - 2V(\pi)\rho/$$
$$[(1+r)^t - 1]\}. \tag{4.69}$$

Comparing this criterion with its riskless analogue of equation (3.19), the effect of risk due to price variability is seen to be the addition of the term involving RSU_{EV} to the marginal opportunity cost of the RHS. This term will be positive so long as $\pi > \Upsilon\pi^{**}(1+r)^t(\partial t/\partial \Upsilon)$. In this case—which must be expected to be the usual one—the effect of risk is to increase marginal cost and hence further shorten the time length t of each run of the process. The same result can be seen in terms of the input decision variables X_i by working directly from equations (4.67), (4.63) and

(4.65) or by expressing equation (4.69) in terms of X_i. Thus we have

$$E(p_y)[\partial T/\partial X_i] = p_i(1+r)^t + \sum(p_i X_i + F)\rho(1+r)^t(\partial t/\partial X_i)$$
$$+ E(\pi^{**})(1+r)^t(\partial t/\partial X_i)$$
$$+ (RSU_{EV})\{2V(p_y)T(\partial T/\partial X_i)\}$$
$$\{\rho/[(1+r)^t - 1]\}\{1 - T\rho(1+r)^t$$
$$(\partial t/\partial T)/[(1+r)^t - 1]\}. \quad (4.70)$$

Comparison of this equation with its risky but timeless analogue of equation (4.51) shows the influence of time added to risk. Conversely, comparison with the riskless analogue of equation (3.24) shows the influence of risk added to time to be the addition of the RHS term involving RSU_{EV}. Note that risk and time effects, just like yield and price uncertainty, are not simply additive—there is also an interactive effect between them.

4.4.5. INTERRELATED YIELD AND PRICE RISKS

Usually yield and price risks will be statistically independent in so far as the farmer's price for his product is not influenced by his own production. However, if climatic effects such as drought and rainfall tend to occur simultaneously across a country or large producing regions rather than in isolated areas, yield and product price will likely be correlated. To illustrate such effects on best operating conditions, we will extend the single decision variable analysis of Section 4.4.1 to allow for non-independence between the probability distributions of yield and product price. For simplicity, we assume these distributions to be normal.

With utility defined as a function of $E(\pi)$ and $V(\pi)$, the dependence between p_y and T is captured by their correlation coefficient ρ_{py}. Equations (4.25) and (4.26) respectively become

$$E(\pi) = E(p_y)E(T) + \rho_{py}[V(p_y)V(T)]^{.5} - p_1 X_1 - F \quad (4.71)$$

$$V(\pi) = [E(p_y)]^2 V(T) + [E(T)]^2 V(p_y) + V(p_y)V(T)$$
$$+ \rho_{py}^2\{\rho_{py}V(p_y)V(T) + 2E(p_y)E(T)[V(p_y)V(T)]^{.5}\}. \quad (4.72)$$

In turn, the criterion of equation (4.30) becomes

$$E(p_y)[dE(Y)/dX_1] = p_1 - \rho_{py}[V(p_y)/V(Y)]^{.5}[dV(Y)/dX_1](0.5)$$
$$+ (RSU_{EV})[\{[E(p_y)]^2 + V(p_y)\}[dV(Y)/dX_1]$$
$$+ 2V(p_y)E(Y)[dE(Y)/dX_1]$$
$$+ \rho_{py}{}^3V(p_y)[dV(Y)/dX_1] + 2\rho_{py}{}^2E(p_y)$$
$$[V(p_y)V(Y)]^{.5}[dE(Y)/dX_1] + \rho_{py}{}^2E(p_y)E(Y)$$
$$[V(p_y)]^{.5}[dV(Y)/dX_1]/[V(Y)]^{.5}]. \qquad (4.73)$$

Comparison with equation (4.30) shows the pervasive effect of correlated yield and price on best operating conditions. Whether it is equivalent to an increase or a decrease in marginal cost depends on the relative size of the second and third groups of terms on the RHS of equation (4.73)—and from these no general implications are apparent. However, it is interesting that for a risk-indifferent decision maker (i.e. one with $RSU_{EV} = 0$), correlation between yield and price is still influential via the second group of terms on the RHS of equation (4.73).

4.4.6. EFFECT OF SKEWNESS

For some decision makers, skewness of the probability distribution of profit—as well as its mean and variance—will be a significant influence. In the moment form of equation (4.13), their utility function is

$$U = f(E(\pi), V(\pi), S(\pi)). \qquad (4.74)$$

Analogously to equation (4.18), best operating conditions imply that the derivatives

$$dU/dX_i = [\partial U/\partial E(\pi)][\partial E(\pi)/\partial X_i] + [\partial U/\partial V(\pi)][\partial V(\pi)/\partial X_i]$$
$$+ [\partial U/\partial S(\pi)][\partial S(\pi)/\partial X_i] \quad (i = 1, 2, \ldots, n) \qquad (4.75)$$

be equal to zero. Hence, akin to equation (4.48), we have

$$0 = \partial E(\pi)/\partial X_i - (RSU_{EV})[\partial V(\pi)/\partial X_i]$$
$$- (RSU_{ES})[\partial S(\pi)/\partial X_i] \qquad (4.76)$$

where RSU_{ES} denotes the rate of substitution in utility of $E(\pi)$ for $S(\pi)$.

If there is both price and yield uncertainty and they are independent with co-moments of zero, profit skewness is approximated by

$$S(\pi) = S(p_y)S(Y) + 3S(p_y)E(Y)V(Y) + 3E(p_y)V(p_y)S(Y)$$
$$+ 6E(p_y)V(p_y)E(Y)V(Y) + S(p_y)[E(Y)]^3$$
$$+ S(Y)[E(p_y)]^3. \qquad (4.77)$$

For simplicity we will assume p_y has a symmetric distribution, i.e. $S(p_y) = 0$. Equation (4.77) then reduces to

$$S(\pi) = 3E(p_y)V(p_y)S(Y) + 6E(p_y)V(p_y)E(Y)V(Y)$$
$$+ S(Y)[E(p_y)]^3. \qquad (4.78)$$

Given an empirical function relating $S(Y)$ to the decision variables X_i, substitution of the derivative for $\partial S(\pi)/\partial X_i$ from equation (4.78) into the skewness-effect term of equation (4.76) gives

$$(RSU_{ES})[\partial S(\pi)/\partial X_i] = (RSU_{ES})[[\partial S(Y)/\partial X_i]\{3E(p_y)V(p_y)$$
$$+ [E(p_y)]^3\} + [\partial V(Y)/\partial X_i]6E(p_y)$$
$$V(p_y)E(Y) + [\partial E(Y)/\partial X_i]\{6E(p_y)$$
$$V(p_y)V(Y)\}]. \qquad (4.79)$$

Whether this skewness effect constitutes an addition or a subtraction to marginal cost depends on the signs of RSU_{ES}, $\partial S(Y)/\partial X_i$ and $\partial V(Y)/\partial X_i$. Intuitively we might expect $\partial S(Y)/\partial X_i$ and $\partial V(Y)/\partial X_i$ to usually be positive. We would also generally expect RSU_{ES} to be negative since most decision makers prefer positive skewness of profit (i.e. tailing of the probability distribution to the right) so that their mean-skewness isoutility curves at a fixed level of variance indicate a need to compensate for lower skewness by a higher mean. However, it is generally unlikely that the marginal cost-reducing effect of positive skewness would ever out-balance the cost-increasing effect of variance for a risk-averse decision maker. In consequence, the marginal cost under uncertainty (i.e. $p_i + RSU_{EV}[\partial V(\pi)/\partial X_i] + RSU_{ES}[\partial S(\pi)/\partial X_i]$) is still most likely to exceed that under certainty (i.e. p_i) and hence induce lower levels of input use.

4.5 Empirical Appraisal under Risk

In the preceding section we have outlined the theoretical analysis relevant to using utility-based decision theory to appraise best operating conditions for risky response. An implicit assumption in our analysis was that the necessary empirical functions or relationships could be specified for real-world analysis. Something now needs to be said at least briefly about these empirical aspects. They relate chiefly to specification of the utility function and probability distributions for profits; the evaluation of expected utility for different choice alternatives; and the use of utility-based but specific utility function-free procedures of stochastic dominance to segregate potential choices into those that are risk efficient and those that are not.

4.5.1. SPECIFICATION OF THE UTILITY FUNCTION

If a decision maker is interested and co-operative, it is usually feasible to elicit his utility function by using his responses to a series of hypothetical questions involving different risky profit possibilities. The basis of the questioning is the continuity axiom of Section 4.3.2. This implies that for any triplet of profits $\pi_1 < \pi_2 < \pi_3$, there exists a probability p such that the utility of π_2 is equal to the utility of the risky choice offering π_1 with a chance of p and π_3 with a chance of $(1 - p)$. Equal utility, of course, implies indifference of choice by the decision maker. Alternatively, the axiom implies that for any risky choice which offers π_1 with probability p and π_3 ($>\pi_1$) with probability $(1 - p)$, there exists a CERTAINTY EQUIVALENT or sure profit π_2 ($>\pi_1$ and $<\pi_3$) whose utility is equal to that of the risky choice. Both these formulations of the axiom imply

$$pU(\pi_1) + (1 - p)U(\pi_3) = U(\pi_2). \qquad (4.80)$$

Using this relationship, questions to determine a decision maker's utility function may be organized in a variety of ways. This is done via a series of linked questions in each of which three of the four quantities p, π_1, π_2 and π_3 are specified and the decision maker chooses the fourth so that equation (4.80) holds true. In other words, alternatives corresponding to the left and right sides of equation (4.80) are posed to the decision maker in such a fashion as to establish the level of the unspecified quantity which induces him to be indifferent between the two sides. Each such indifference relationship establishes a point in $[\pi, U(\pi)]$ space. Given a series of such points, the utility curve $U(\pi)$ can be drawn as in Fig. 4.1 or statistical estimation can be used to obtain some appropriate algebraic estimate $U = f(\pi)$ of the utility function. For purposes of subsequent analysis, the reading of utility values from a graph of the utility function will often be sufficient. Other times, it will be more convenient to have available an algebraic specification of the utility function.

One reasonable procedure for eliciting the required set of $[\pi, U(\pi)]$ points is that known as the modified von Neumann–Morgenstern method. In terms of equation (4.80), it uses $p = 0.5$ and questions are posed to determine the certainty equivalent π_2 of the fifty-fifty or even-chance prospect involving given values of π_1 and π_3. The trick is to choose the π_1 and π_3 values in such a way that a linked series of questions are generated to span the relevant range of profit possibilities (and hence utilities) that confront the decision maker in his real-world activities. Suppose the relevant range of profit is from a low of π_0 to a high of π_{100}—the relevance of the subscripts will become apparent below. Since the utility scale is arbitrary, we can set $U(\pi_0) = 0$ and $U(\pi_{100}) = 100$. Questions are then posed to determine the certainty equivalent π_{50} such that

$$0.5U(\pi_0) + 0.5U(\pi_{100}) = U(\pi_{50}).$$

Since $U(\pi_0) = 0$ and $U(\pi_{100}) = 100$, it follows that $U(\pi_{50}) = 50$. We now have three points on the utility curve. These are $(\pi_0, 0)$, $(\pi_{50}, 50)$ and $(\pi_{100}, 100)$. The next two points are obtained by repeating the procedure to find π_{25} and π_{75} respectively such that

$$0.5U(\pi_0) + 0.5U(\pi_{50}) = U(\pi_{25})$$

and

$$0.5U(\pi_{50}) + 0.5U(\pi_{100}) = U(\pi_{75})$$

so that $U(\pi_{25}) = (0.5)(0) + (0.5)(50) = 25$ and $U(\pi_{75}) = (0.5)(50) + (0.5)(100) = 75$. The procedure of bisecting the already established utility intervals is repeated in analogous linked fashion to obtain as many further points as desired. Usually seven to nine will be sufficient and checks can be made for consistency, repeating the questions if need be. Thus the same value for π_{50} as a certainty equivalent should result for the risky choice with $[0\cdot5U(\pi_{25}) + 0\cdot5U(\pi_{75})]$ as from $[0\cdot5U(\pi_0) + 0\cdot5U(\pi_{100})]$.

For purposes of risky response analysis, it will usually be desirable to have an algebraic estimate of the utility function. This is because risky response involves input decision variables which are continuous rather than discrete and which, at each level, have some continuous distribution of possible yields rather than some small array of discrete yield possibilities. Statistical procedures therefore need to be used to estimate some algebraic fit of $U = f(\pi)$ to the set of elicited points defining the curve. This can be done by least-squares regression using, for example, the quadratic, logarithmic and power functions of equations (4.10), (4.11) and (4.12) respectively. The one which fits best (on the basis of subjective assessment) may then be chosen for further use. Note, for example, that if a quadratic function is fitted it will be of the form

$$U = a_0 + a_1\pi + a_2\pi^2 \quad (4.81)$$

which, by virtue of utility only being defined up to a positive linear transformation, can be transformed to the simpler form of equation (4.10) where, in terms of equation (4.6), $\alpha_1 = 1/a_1$ and $\alpha_2 = -a_0/a_1$ so that $b = a_2/a_1$.

4.5.2. EVALUATION OF BEST OPERATING CONDITIONS

Given an algebraic form $U = f(\pi)$ for the utility function, the utility objective function (4.9) can be specified as

$$U = \int_{-\infty}^{\infty} f(\pi) h(\pi | X_1, \ldots, X_n) d(\pi). \quad (4.82)$$

Evaluation of alternatives by direct maximization of this function, however, will only be simple and convenient if the probability distribution $h(\pi | X_1, \ldots, X_n)$ is of a very simple type (such as discrete, rectangular or triangular).

Usually, evaluation will be more conveniently carried out by decomposition procedures based on the mean $E(\pi)$ and the moments M_k about the mean of the probability distribution of profit. Two approaches are pertinent—one of a discrete enumerative nature, the other of a continuous and analytical orientation. The latter is relevant if the decision variables are continuous and thus have an infinite array of possible levels. This is usually the case in response processes, as for example with fertilizer in crop production. However, as noted below, it may sometimes be convenient to treat such continuous variables as if they had only a limited possible number of discrete levels. The continuous analytical approach is the one we have already used throughout Section 4.4. Its essence is the setting of the derivatives dU/dX_i equal to zero and solving for the set of utility maximizing X_i values. As shown by the numerical example of Section 4.4.1, the data requirements of the continuous or analytical method consists of knowledge of the algebraic form of the utility function and of the mean and relevant moments about the mean of each of the profit distributions. As pertinent, these profit probability parameters should be expressed as functions of the decision variables.

The discrete enumerative approach applies if the decision variables are discrete or if it is judged that they may be reasonably treated as if they were discrete. For example, fertilizer level might be appraised not as a continuous factor but as a discrete variable with units of, say, 10 kilograms. The essence of the discrete enumerative approach is to calculate the expected utility of each discrete choice possibility. The alternative which yields the greatest utility is optimal. To calculate the utility of each potential choice, a very convenient moment-based procedure is available. This procedure derives from the convenient (if not extraordinary) fact that the utility of a risky choice is equal to its expected utility. Based on a Taylor series expansion of the expected utility function (4.82) taken about the mean profit level, this moment method expresses expected utility—and hence the utility objective function—in the form

$$U = U[E(\pi)] + U_2[E(\pi)]M_2(\pi)/2! + U_3[E(\pi)]M_3(\pi)/3! + \ldots,$$
(4.83)

where $U[E(\pi)]$ is the value of the utility function at the mean profit level; $U_k[E(\pi)]$ is the kth derivative $d^kU/d\pi^k$ of the utility function evaluated at $E(\pi)$; and $M_k(\pi)$ is the kth moment about the mean of the probability distribution of profit, i.e. $M_k(\pi) = E[\pi - E(\pi)]^k$.

If the utility function is quadratic, just the first two terms (involving profit mean and variance) of equation (4.83) provide an exact assessment of utility since

$$U = U[E(\pi)] + 2bM_2(\pi)/2! \tag{4.84}$$

$$= E(\pi) + b[E(\pi)]^2 + bV(\pi) \tag{4.85}$$

$$= E(\pi + b\pi^2). \tag{4.86}$$

Likewise the first three terms involving profit mean, variance and skewness $[M_3(\pi) \equiv S(\pi)]$ are exact for the cubic function

$$U = E(b_1\pi + b_2\pi^2 + b_3\pi^3), \quad b_2^2 < 3b_1b_3, \ b_1 > 0, \tag{4.87}$$

$$= b_1 E(\pi) + b_2 \{V(\pi) + [E(\pi)]^2\} + b_3 \{S(\pi) + 3E(\pi)V(\pi) + [E(\pi)]^3\}. \tag{4.88}$$

For the logarithmic and power utility functions of equations (4.11) and (4.12), the moment method is only approximate though usually adequate. The logarithmic function becomes

$$U = \log_e[W + E(\pi)] - [W + E(\pi)]^{-2}V(\pi)/2 + [W + E(\pi)]^{-3}S(\pi)/3, \tag{4.89a}$$

and the power function yields

$$U = [W + E(\pi)]^c + c(c-1)[W + E(\pi)]^{c-2}V(\pi)/2 + c(c-1)(c-2)[W + E(\pi)]^{c-3}S(\pi)/6. \tag{4.89b}$$

The data requirements of the discrete enumerative method of appraisal are exactly the same as for the continuous analytical approach: first, knowledge of the algebraic form of the utility function; and, second, knowledge of the mean and variance, and, if required, the skewness of each probability distribution of profit expressed, if relevant, as functions of the discrete decision variables.

To illustrate the discrete approach, suppose we are happy to evaluate nitrogen use in the empirical example of Section 4.4.1 by assuming discrete N units of 10 kilograms. Use of equation (4.84) to calculate utility values with both price and yield risk gives:

N (kg/ha):	0	50	60	70	80	90	100
Utility:	9088	9882	9926	9945	9943	9923	9888

The discrete approximation thus suggests an optimal N of 70 as against 74 kilograms per hectare by the continuous or analytical procedure.

4.5.3. SPECIFICATION OF THE PROBABILITY DISTRIBUTIONS

A number of aspects of the probability distribution appearing in the utility objective function of equations (4.8) or (4.9) have already been noted. First, because of the one-to-one mapping from π to $U(\pi)$ via the utility function, the probability distribution of profit corresponds directly to the probability distribution of possible utility outcomes for a particular choice. Second, for each possible setting of the decision variables there will be a corresponding probability distribution of profit. The risky response problem is to choose the most attractive of these alternative distributions. Third, each profit distribution is a function of whatever random or risky variables enter the profit equation (4.2). In general, the risky variables will be product price p_y and yield Y corresponding respectively to price risk and yield risk. Fourth, if both price and yield risk are relevant, each profit probability distribution will correspond to a joint distribution of p_y and Y. Generally product price and yield will be independent for an individual decisionmaker. If they are not, their dependence should be allowed for as in equations (4.71) and (4.72) and via the co-third moments if skewness is relevant to utility appraisal. Fifth, if it exists, product price risk will be pertinent through its influence on the gross revenue term $p_y Y$ of the profit function (4.2). Sixth, yield risk arising from uncertain uncontrolled input factors will only be relevant if there is interaction between these uncertain factors and the decision variables. If the probability distribution of yield is not conditional on the decision variables, yield risk has no influence on best operating conditions. Seventh, only if the probability distribution of profit or its source distributions of price and yield are of simple form— e.g., discrete, rectangular or triangular—will direct appraisal via the algebraic expression for the distribution as per equation (4.82) be feasible. Usually it will be far more convenient, if not essential, to work with the mean, variance and perhaps skewness parameters of the relevant probability distributions. Eighth, by virtue of the expected utility theorem, subjective probability distributions can always be specified for the uncertain outcomes of relevance. These distributions should be those judged appropriate by the decision maker. They may, of course, be influenced to a greater or lesser extent by objective historical data.

Empirically, the major probability aspect is to determine the mean $E(\pi)$, the variance $V(\pi)$ and, if relevant, the skewness $S(\pi)$ of profit for each choice alternative. If sufficient guideline data of a historical nature are available on gross returns $p_y Y$, estimation may proceed directly. What are needed, if both price and yield risk are pertinent, are time-series data conditional on the controlled inputs $X_1, \ldots, X_n$. When this approach is possible, it has the advantage of avoiding the need to account for price and yield risks separately. Most often, however, while data on $p_y Y$ may be available, its conditional association on $X_1, \ldots, X_n$ will not be known and the approach cannot be used.

Determination of $E(p_y)$, $V(p_y)$ and $S(p_y)$ usually presents no problems. Often time-series data will be known as a guide. If not, unguided subjective (but none the less the only valid) estimates can be made. Moreover since p_y is typically independent of the decision variables, a single value of each parameter $E(p_y)$, $V(p_y)$ and $S(p_y)$ serves for appraisal of the complete array of input choice possibilities.

In concept, the probability distribution of yield is a little more difficult. At its most complicated, it may be conditional not just on the decision variables $X_1, \ldots, X_n$ of response function (4.1) but also on the predetermined variables $X_{n+1}, \ldots, X_k$. This is the usual case with, for example, interaction between fertilizer decision variables, uncertain climatic variables and predetermined soil fertility factors. At its simplest, the yield distribution may only be conditioned by the decision variables. One possibility would be to attempt to specify the yield distribution in relation to the joint distribution of the uncertain variables, conditional on the relevant array of known (decision and predetermined) variables. Never, however, will the full array of uncertain and predetermined variables be known. At best all that can be done is to pick out a few obviously important variables such as, for example, the soil's pH, nitrogen, phosphorus and potash status, rainfall and temperature. Even this limited array might be further composited into some one or two surrogate indices. Such an analytical approach to the yield distribution is not to be recommended. Because of the impossibility of capturing all the relevant factors, it is sure to lead to a relatively poor estimate of yield risk.

136 THE ANALYSIS OF RESPONSE IN CROP AND LIVESTOCK PRODUCTION

The preferred alternative approach to assessing the probability distribution of yield is to proceed directly to the fitting of empirical functions

$$E(Y) = f_E(X_1, \ldots, X_n), \quad (4.90)$$

$$V(Y) = f_V(X_1, \ldots, X_n), \quad (4.91)$$

$$S(Y) = f_S(X_1, \ldots, X_n) \quad (4.92)$$

which relate the parameters of the yield distribution to the level of the decision variables. Such an approach, which might be described as gross rather than analytical, is the one we have followed in the example of Section 4.4.1 above where equations (4.39) and (4.40) correspond respectively to equations (4.90) and (4.91). The data required to fit such empirical relationships are a series of observations on the probability parameters and their corresponding levels of the decision variables $X_1, \ldots, X_n$. In making the estimates, it is assumed that variations in the uncertain variables occur across the $X_1, \ldots, X_n$ combinations, thereby reflecting the sources of yield uncertainty. If observations are available on any of the predetermined but uncontrolled variables $X_{n+1}, \ldots X_k$, the empirical functions (4.90), (4.91) and, if relevant, (4.92) should be extended to include these predetermined variables as well as the decision variables. In this way the gross approach will capture as many as possible of the factors conditioning the parameters of the probability distribution of yield.

So far we have shown the importance of having estimates of such parameters as the mean, variance and skewness of Y and p_y (and of π if we are not working through Y and p_y), but we have not shown how to obtain these estimates. Two types of situation may be distinguished—one with ample data, the other with sparse data, i.e. say less than 10 observations.

With 10 or more observations, the standard moment-estimation formulae may be used to provide guideline estimates. Thus for the set of observations $x_1, x_2, \ldots, x_n$ on the random variable x (corresponding to p_y, Y or π):

$$E(x) = \sum x_i / n \quad (4.93)$$

$$V(x) = [\sum x_i^2 - (\sum x_i)^2/n]/(n-1) \quad (4.94)$$

$$S(x) = [n\sum x_i^3 - 3\sum x_i \sum x_i^2 + 2(\sum x_i)^3/n]/(n-1)(n-2). \quad (4.95)$$

For example, in terms of Y the above procedure would imply we had 10 or more observations on Y at each of some sufficient number of combinations of the decision variables $X_1, \ldots, X_n$. For each of these input combinations (usually experimental treatments), equations (4.93), (4.94) and (4.95) could be applied and then least-squares regression used to fit equations (4.90), (4.91) and (4.92).

With less than 10 observations, sparse-data procedures are relevant. These are based on the fact that if we have N observations on a random variable and we arrange them in ascending order of size, the expected fraction of all possible values of the random variable falling below the Kth observation is $K/(N+1)$. Thus the Kth ordered observation is an estimate of the $K/(N+1)$ fractile of the random variable's distribution. For example, suppose we have three years of observations on p_y, say values of 6, 3 and 8. In ordered form they are 3, 6 and 8, respectively giving estimates of the 1/4, 2/4 and 3/4 fractiles of the p_y distribution, i.e. there is a probability of 0·25, 0·5 and 0·75 that a random drawing of p_y will lie below 3, 6 and 8 respectively. Though barely sufficient, with this little information we can sketch the cumulative distribution of p_y. This sketch, of course, will be subjective and be guided by other relevant information—e.g., we might assume the lowest possible value for p_y is zero, and the highest 11. Figure 4.5 illustrates the procedure using the above data.

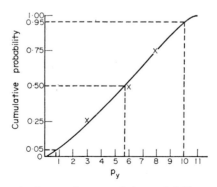

FIG. 4.5. Example of sparse-data cumulative probability curve for use in moment estimation.

138   THE ANALYSIS OF RESPONSE IN CROP AND LIVESTOCK PRODUCTION

Given a graph of the cumulative distribution, moments can be calculated by breaking the curve into, say, 20 equal probability intervals and calculating the moments using the probability elements (0·05 if 20 intervals) and associated values of the random variable at the midpoints of the probability intervals. Thus $E(x) = \sum x_i P_i$ where x_i is the midpoint value and P_i the probability interval. Likewise, $M_k(x) = \sum [x_i - E(x)]^k P_i$. More conveniently, however, a reasonable approximation formula is available for the mean and, to a less adequate degree, for the variance. These are

$$E(x) \simeq F_{.5} + (0\cdot 185)(F_{.95} + F_{.05} - 2F_{.5}) \qquad (4.96)$$
$$V(x) \simeq [(F_{.95} - F_{.05})/3\cdot 25]^2 \qquad (4.97)$$

where the fractiles $F_{.05}$, $F_{.5}$ and $F_{.95}$ are read from the sparse-data cumulative probability curve. Reading from Fig. 4.5, for our example we have $F_{.05} = 0\cdot 73, F_{.5} = 5\cdot 7$ and $F_{.95} = 10$. Hence $E(p_y) \simeq 5\cdot 58$ and $V(p_y) \simeq 8\cdot 14$.

As in the case with 10 or more observations, once the required set of moment estimates have been calculated for each available combination of the variables, equations (4.90), (4.91) and, if required, (4.92) can be fitted by least-squares regression.

4.6 Stochastic Dominance Analysis

Our preceding analysis of risky response has assumed knowledge of the decision maker's utility function. More general appraisal is possible by means of the rules of stochastic dominance. In contrast to the identification of the risk–optimal input combination that is possible with knowledge of the utility function, stochastic dominance analysis only identifies sets of risk-efficient operating conditions. The essence of the method lies in the comparison of entire probability distributions, each of which corresponds to the array of possible profits associated with a particular choice alternative. The comparisons are made by checking whether or not various cumulative curves of the distributions cross one another. Figure 4.6 provides a simple example where curves A, B and C might respectively correspond to the alternative choices of using (N, P) fertilizer combinations of, say, $(0, 0)$, $(40, 20)$ and $(100, 40)$ in kilograms per hectare.

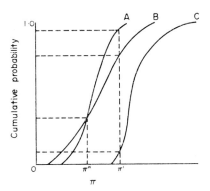

Fig. 4.6. Illustration of the principle of stochastic dominance analysis.

Because the cumulative probability distribution depicted by curve C lies everywhere to the right of curves A and B, alternative C, i.e. $(N, P) = (100, 40)$, would be preferred over alternatives A and B by all decision makers who prefer more profit to less. The reason is that for any profit level such as π' that could occur with any of the alternatives A, B or C, alternative C offers the smallest (greatest) probability that actual profit will fall below (above) this level. So C is risk efficient compared to A and B. Now compare A and B. For π values below π'', A is always best; but above π'', B is always best. Hence we cannot make an unqualified statement about the relative merits of A and B.

The above example is, in fact, an illustration of the first rule of stochastic dominance known as first-degree stochastic dominance. There are also second and third-degree rules. Their formal definition is as follows.

Consider two probability distributions $h(\pi)$ and $g(\pi)$ defined over the profit range $\pi = a$ to $\pi = b$. Suppose $h(\pi)$ relates to an alternative labelled H and $g(\pi)$ to another labelled G. Given $h(\pi)$ and $g(\pi)$, we can calculate the quantities

$$H_1(R) = \int_a^R h(\pi)\, d\pi \qquad G_1(R) = \int_a^R g(\pi)\, d\pi \qquad (4.98)$$

$$H_2(R) = \int_a^R H_1(\pi)\, d\pi \qquad G_2(R) = \int_a^R G_1(\pi)\, d\pi \qquad (4.99)$$

$$H_3(R) = \int_a^R H_2(\pi)\, d\pi \qquad G_3(R) = \int_a^R G_2(\pi)\, d\pi \qquad (4.100)$$

for R values from a to b. These quantities are the first, second and third-order cumulatives of $h(\pi)$ and $g(\pi)$. The quantities $H_1(R)$ and $G_1(R)$ for R from a to b are, of course, familiar as the points plotting out the cumulative probability distribution for $h(\pi)$ and $g(\pi)$ respectively. In similar fashion equations (4.99) and (4.100) can respectively be used to generate values plotting out the second- and third-order cumulative curves. These first-, second- and third-order cumulative curves are useful in comprehending the following rules of stochastic dominance.

FIRST-DEGREE STOCHASTIC DOMINANCE of H over G prevails if

$$H_1(R) \leq G_1(R) \qquad (4.101)$$

for all R in the range $[a, b]$ with $H_1(R) < G_1(R)$ for at least one value of R. Compared to G, H is said to be FIRST-DEGREE RISK EFFICIENT.

SECOND-DEGREE STOCHASTIC DOMINANCE of H over G prevails if

$$H_2(R) \leq G_2(R) \qquad (4.102)$$

for all R in the range $[a, b]$ with $H_2(R) < G_2(R)$ for at least one value of R. Compared to G, H is said to be SECOND-DEGREE RISK EFFICIENT.

THIRD-DEGREE STOCHASTIC DOMINANCE of H over G prevails if

$$H_3(R) \leq G_3(R) \qquad (4.103)$$

for all R in the range $[a, b]$ with $H_3(R) < G_3(R)$ for at least one value of R. Compared to G, H is said to be THIRD-DEGREE RISK EFFICIENT.

These rules relate to decision makers' utility functions in the following manner. If H is first-degree risk efficient but G is not, H will be preferred over G by any decision maker whose marginal utility for profit, $dU/d\pi$, is positive, i.e. so long as the utility function is monotonically increasing. Since this requirement merely corresponds to preferring more profit to less, it is obviously reasonable.

If neither H nor G are first-degree risk efficient but H is second-degree risk efficient over G, then H will be preferred to G by any decision maker whose utility function has $dU/d\pi$ positive and $d^2U/d\pi^2$ negative, i.e. by any risk-averse decision maker.

If neither H nor G are first- or second-degree risk efficient but H is third-degree risk efficient over G, then H will be preferred to G by any decision maker whose utility function has $dU/d\pi$ positive, $d^2U/d\pi^2$ negative and $d^3U/d\pi^3$ positive, i.e. more or less so long as he is decreasingly risk averse as his wealth increases.

Application of the stochastic dominance rules is illustrated diagrammatically in Fig. 4.7 where the cumulative curves based on the quantities calculated from equations (4.98), (4.99) and (4.100) have been plotted as relevant for six alternatives labelled I to VI. Figure 4.7(a) allows inspection for first-degree stochastic dominance. On this basis we can eliminate alternatives I, II and III (I and II are dominated by III to VI, and III is dominated by IV and V). The first-degree risk efficient set of alternatives thus consists of IV, V and VI. The second-order cumulatives of these three are plotted in Fig. 4.7(b). Inspection shows VI to be dominated by IV so that the second-degree risk efficient set consists of alternatives IV and V. Plotting the third-order cumulatives of IV and V in Fig. 4.7(c) indicates that the set of third-order risk efficient alternatives consists only of IV. Note that in graphing the second- and third-order cumulatives, each curve is plotted up to the highest value of R taken by any of the first-order cumulatives.

Just as individual utility appraisal has the disadvantage of requiring knowledge of the utility function but the advantage of only needing to know the first few moments of each profit distribution, stochastic dominance analysis has the advantage of not needing full information about the utility function but the disadvantage of needing full specification of each alternative's profit distribution. Too, stochastic dominance analysis loses its generality unless the decision making audience to whom the analysis is addressed hold subjective probability distributions well-matched by those used in the appraisal of risk efficiency.

Empirically, application of the rules of stochastic dominance can be a demanding task. It necessarily involves, first, the "sketching" (visual or by computer) of the relevant probability distributions (perhaps using sparse-data procedures) and, second, numerical methods of integration to enable appraisal (visual or by computer) of the first-, second- and third-order cumulatives. Also, because the decision variables involved in response analysis are continuous, approximation via many discrete combinations of levels is necessary for thorough appraisal. For example, stochastic dominance appraisal of possible (N, P) fertilizer decisions for a particular crop in a particular region might necessitate considering 11

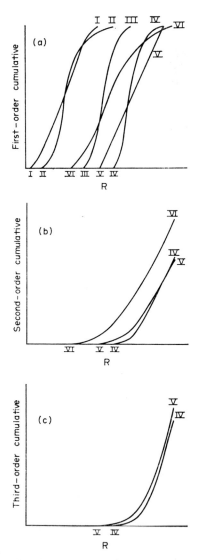

Fig. 4.7. Example of risk efficiency analysis using the rules of (a) first-degree stochastic dominance, (b) second-degree stochastic dominance and (c) third-degree stochastic dominance.

levels of N (from say 0 to 200 kilograms per hectare in steps of 20 kilograms) in combination with each of 11 levels of P (from say 0 to 80 in steps of 8 kilograms per hectare). All told there would thus be 121 profit probability distributions to be specified and appraised. The results of such appraisal might, for example, be as illustrated in Table 4·2. Those combinations with a zero entry in the table are risk inefficient; those with a 1, 2 or 3 entry are risk efficient of the first, second or third degree respectively. Note that an entry of 3 implies 2 and 1 also; and an entry of 2 also implies 1.

TABLE 4.2

HYPOTHETICAL ILLUSTRATION OF THE RESULTS OF STOCHASTIC DOMINANCE
APPRAISAL OF ALTERNATIVE FERTILIZER COMBINATIONS

Nitrogen (kg/ha)	Phosphorus (kg/ka)									
	0	8	16	24	32	40	48	56	64	72	80
0	0	0	1	2	3	3	3	3	3	3	3
20	0	0	0	0	1	3	3	3	3	3	2
40	0	0	0	0	1	2	3	3	3	2	1
60	0	0	0	0	1	1	3	3	2	1	1
80	0	0	0	0	1	1	3	2	1	1	1
100	0	0	0	0	0	1	2	2	1	1	1
120	0	0	0	0	0	1	1	1	1	1	1
140	0	0	0	0	0	0	1	1	1	1	1
160	0	0	0	0	0	0	1	1	1	1	1
180	0	0	0	0	0	0	0	1	1	1	1
200	0	0	0	0	0	0	0	1	1	1	1

4.7 Further Reading

The decision theory approach to best operating conditions for risky response, along with extensive discussion of the concept and elicitation of both utility functions and subjective probabilities, is presented by Anderson, Dillon and Hardaker (1976, chs. 2, 4 and 6). Some of the same ground is covered by Dillon (1971) and Halter and Dean (1971). More wide-ranging mathematical treatment, including consideration of second-order conditions for utility maximization, is to be found in Magnusson (1969). Expected utility analysis within the general setting of microeconomics is surveyed by McCall (1971).

While we have used the terms risk and uncertainty interchangeably to denote probabilistic decision situations, the term risk is also commonly used to denote the probability of loss, e.g. de Janvry (1972a), and in a technical statistical sense to denote a mean-preserving spread of a probability distribution, e.g. Diamond and Stiglitz (1974). Acceptance of the expected utility theorem, of course, implies discarding the old distinction between risk (where frequency data are available) and uncertainty (where no frequencies are known).

The use of expected profit, safety-first and game theoretic criteria to appraise risky response is variously exemplified by Carruthers and Donaldson (1971), Cone (1974-5), de Janvry (1972a), Chisholm (1965), Doll (1972), Heady, Pesek and McCarthy (1963), Roumasset (1974 and 1976), Swanson and Tyner (1965) and Walker, Heady and Pesek (1964).

Of necessity, we have skipped over a number of topics without proof or elaboration. An easily comprehended proof of the expected utility theorem is given by Dyckman, Smidt and McAdams (1969). Taylor series expansion of the utility and expected utility function is elaborated by Dillon (1971). For the construction of sparse-data probability curves, see Anderson (1973 and 1974a) and Anderson, Dillon and Hardaker (1976, chs. 2 and 6). Chapter 8 of the latter reference also discusses the difficulty of utility appraisal involving risky time sequences. The theory of first and second-degree stochastic dominance is presented by Hadar and Russell (1969), and for the third-degree by Whitmore (1970). All three rules are covered by Hadar and Russell (1974). Elaboration and empirical examples of stochastic dominance and risk efficiency analysis in the context of agricultural response are given by Anderson (1974b) and Anderson, Dillon and Hardaker (1976, ch. 9). The latter reference also contains a computer programme for stochastic dominance analysis.

Apart from the examples presented by Anderson (1973, 1974a and b), Anderson, Dillon and Hardaker (1976, ch. 6) and Halter and Dean (1971), there appears as yet to be no full applications of utility-based risky response analysis. Closest are the sheep stocking rate work of McArthur (1970) and McArthur and Dillon (1971), the fertilizer-solar energy study of Montaño and Barker (1970) with rice, the broiler study —notable for its incorporation of both time and price uncertainty in an expected profit framework—of Hochman and Lee (1972), and the farm efficiency study of Dillon and Anderson (1971). Of close interest is the analysis of risk and farmer learning (i.e. adjustment of subjective probabilities) in new crop response processes by Hiebert (1974). Studies involving the mean and variance of return include those of Battese and Fuller (1972), Battese, Fuller and Shrader (1972), Colyer (1969), Ryan and Perrin (1973) and Tollini and Seagraves (1970). Also relevant as steps along the way to the more comprehensive decision theory approach are the earlier contributions of Chisholm (1965), Dowdle (1962), Fuller (1965), Halter (1963) and Smith and Parks (1967).

Examples of the influence of inter-period climatic variability on response are to be found in Battese and Fuller (1972), Brown and Merril (1958), Colwell (1973a), Colyer and Kroth (1970), Doll (1972), Englestad and Parks (1971), Fuller (1965), Hjelm (1962), I.R.R.I. (1969, 1970 and 1972), Pequignot and Recamier (1962), Russell (1968a) and Voss, Hanway and Fuller (1970). The incorporation of climatic and other environmental variables in the response function has been used or investigated by Byerlee and Anderson (1969), de Janvry (1972a), Dean *et al.* (1972), Englestad and Doll (1961), Heady *et al.* (1964a and b), Knetsch (1959), McArthur and Dillon (1971), Montaño and Barker (1970), Oury (1965), Parks and Knetsch (1959 and 1960), Roumasset (1974), Russell (1968c), Ryan and Perrin (1973 and 1974), and Tollini and Seagraves (1970). In an interesting study, Mandac (1974) has fitted rice response functions involving nitrogen and phosphorus fertilizer, degree and time of water stress, solar energy, and soil, disease, insect and weed factors.

Estimation of empirical relationships between input decision variables and the variance of the yield distribution has been carried out by Battese and Fuller (1972), Doll (1972), and Fuller (1965), while Anderson (1973) and Day (1965) have also estimated analogous functions for skewness. More general discussion of climate in relation to response—with differing conclusions as to how best to incorporate climatic uncertainty—is presented by Anderson (1971), Doll (1967), Pesek (1973), Ryan (1972), and Shaw (1964). The use of weather predictions in a decision theoretic response framework is explored by Byerlee and Anderson (1969) and Doll (1971). The modelling of crop response to weather factors is illustrated by Denmead and Shaw (1962), Nix and Fitzpatrick (1969), Oury (1965), and Reid and Thomas (1973).

Appraisal of uncertainties due to pests and disease, as well as climate,

146   THE ANALYSIS OF RESPONSE IN CROP AND LIVESTOCK PRODUCTION

is given by Roumasset (1974) in terms of a "normal" response function which is adjusted multiplicatively by a probabilistic "damage factor".

Further discussion of response variability over time and space, with emphasis on research implications, is given in Section 5.4 below.

Anderson, Dillon and Hardaker (1976), Blandford and Currie (1975), and Hazell and Scandizzo (1975) have considered some of the implications for agricultural policy arising respectively from yield and price uncertainty.

4.8 Exercises

4.8.1. Solve the numerical example of Section 4.4.1 if the farmer's utility function is as in equation (4.11) with $W = \$100,000$. What if also $\rho_{py} = -0.4$?

4.8.2. Is price uncertainty a relevant consideration if the decision variables $X_1, \ldots, X_n$ and the uncertain variables $X_{k+1}, \ldots, X_m$ do not interact?

4.8.3. Determine best operating conditions for the crop response process specified by

$$E(Y) = 5000 + 32N + 40P - 0.08N^2 - 0.03P^2 + 0.1NP$$

$$V(Y) = (600 + 20N + 10P + 0.2N^2 - 0.1NP)^2$$

where Y, N and P are in kilograms per hectare, $p_P = 0.3$ per kilogram, $p_N = 0.4$ per kilogram, and the only four observations available on p_y indicate values of 0.05, 0.07, 0.04 and 0.09 per kilogram. Assume fixed costs are $80 per hectare and that 100 hectares of the crop are to be grown by a farmer whose utility function is $U = (150,000 + \pi)^{.6}$.

4.8.4. Develop criteria for best operating conditions of a response process in which there is uncertainty about the prices of the multiple decision variables but no uncertainty about yield or product price.

4.8.5. Derive best operating criteria for a multiple response situation where total outlay on the decision variables is constrained and each process exhibits price but not yield uncertainty.

4.8.6. Extend the analysis of Section 3.6.2 to incorporate yield risk.

4.8.7. On the basis of the modified von Neumann–Morgenstern method, postulate a series of answers and questions corresponding to the utility curve of Fig. 4.1(a).

4.8.8. Construct three probability distributions such that two are stochastically dominant at the second-degree and one would be preferred by a decreasingly risk-averse decision maker.

4.8.9. The mean and variance of a grazier's annual net income might be specified as follows:

$$E(\pi) = H\{S([M - dS]R - C) - F\},$$
$$V(\pi) = (dAS^2R)^2\sigma^2$$

where H is farm area, S is sheep stocking rate per hectare, M is maximum wool cut per sheep per year, d is the reduction in annual wool cut per sheep per unit increase in S, R is wool price per kilogram, C is variable cost per sheep per year, F is annual fixed cost per hectare and σ^2 is the variance of a climatic index. Deduce the impact on optimal stocking rate of one-at-a-time increases in H, C, F, M and R. What if there were no climatic uncertainty but R had a variance of $V(R)$?

4.8.10. If p_y and P_N are respectively $0.04 and $0.30 per kilogram, apply risk efficiency analysis to evaluate the use of nitrogen given the following set of data:

Nitrogen per hectare	Yield observations in kilograms per hectare				
0	200	1200	2300	4100	1400
20	2000	4900	3400	6600	4300
40	6700	5900	5200	7200	4400
60	4300	7400	7000	9100	6000
80	7200	9600	3700	7800	10000
100	5100	9300	11400	7900	6400
120	8400	9100	4100	10400	10000
140	9200	9900	2800	5500	10700
160	8300	9600	7700	1900	8300

4.8.11. Given the data of Exercise 4.8.10, what level of nitrogen per hectare would be best for a farmer planning to grow 100 hectares of crop with a fixed cost of $100 per hectare if his utility function were:

(i) $U = \pi$.
(ii) $U = \pi - 0{\cdot}00002\pi^2$.
(iii) $U = \pi - 0{\cdot}00002\pi^2 + 0{\cdot}000001\pi^3$.

4.8.12. What if in Exercise 4.8.10 there were a fifty-fifty chance that p_y would be $0·02 or $0·06 per kilogram?

4.8.13. Extend the analysis of Section 3.9.2 under the additional assumption that both product price and the per cent protein in the ration are uncertain. Would it simplify the analysis to assume each distribution is normal?

4.8.14. What implications do you see for the design of crop and livestock response experiments arising from a recognition of the influence of risk?

CHAPTER 5

Difficulties in Response Research

5.1 Introduction

A great many problems arise as soon as we enter the real world of empirical crop and livestock response research and extension. Inevitably, theoretical ideals have to bend to pragmatic practicalities. Over and above such questions as carry-over effects and risk which we have already discussed, the most important difficulties are those concerned with the *interrelated* questions of:

(1) the statistical design of response experiments;
(2) the statistical estimation of response functions and related quantities;
(3) response variability over space and time;
(4) the choice of response model;
(5) the economics of response research;
(6) discrepancies between response under experimental conditions and farm conditions; and
(7) the making of recommendations to farmers.

To do them proper justice, each of these problems of research and extension could easily form the subject of a separate chapter, if not more. That, however, would take us well beyond the aim of an introduction to the principles of crop and livestock response analysis. Accordingly, we round off this introduction by noting some of the more important aspects of these problems and the work that has been done on them. An earlier view of many of these same difficulties (and others since resolved) is given by Baum *et al.* (1956 and 1957), Heady and Dillon (1961, chs. 5, 6, 7) and Hoglund (1959). With luck, the expanding interest being shown in these problem areas of response research and its application will lead to their amelioration if not solution.

5.2 Experimental Design

To enable proper statistical estimation and subsequent economic analysis of the response function, response experiments must be based on an appropriate statistical design. The principles of response analysis, just like the principles of statistical analysis, can be best applied to data that is generated meaningfully. In this regard the response analyst is in exactly the same situation as the statistician—if he does not participate in the planning stage of an experiment, it is more than likely that he will be unable to analyse the experiment adequately. Farmers and others might then justifiably claim that adaptive research resources are being used inefficiently.

A large literature exists on criteria relevant to the choice of an appropriate response surface design and the types of experimental design that are available. Textbook treatment at varying degrees of sophistication and completeness is given by, for example, Cochran and Cox (1957, chs. 1, 2, and 8A), Johnson and Leone (1964, ch. 17), Mendenhall (1968, ch. 10), Myers (1971), and Snedecor and Cochran (1967). Discussion from an applied, practical orientation is given by Anderson (1971), Anderson and Nelson (1975), and Boyd (1973) in particular, and also by Anderson and Dillon (1969), Balaam (1975), Bicking and Gillespie (1963), Bofinger and Wheeler (1975), Bradley and Hunter (1958), Colwell (1976, ch. 10), Dillon (1966), Heady and Dillon (1961, chs. 5 and 7), Hoffnar and Johnson (1966), Kenworthy (1963), O.E.C.D. (1969b), and Williams and Baker (1968). More formal statistical treatment, though generally with an orientation to industrial experimentation, is provided by Box and Draper (1971 and 1974), Cochran (1973), Elandt (1963), Federer and Raghavarao (1975), Hill and Hunter (1974), Kupper and Meydrech (1974), Lucas (1974) and—in review—St. John and Draper (1975) and Hill and Hunter (1966).

In general, the more purposively and systematically an experiment covers the relevant region of the response surface with treatments, the better the response function can be estimated (and analysed). Relative to the traditional "Yes or No" response experiment using functional analysis of variance to ascertain if there are "significant differences" between some few levels of a factor, response function experiments aimed at ascertaining best operating conditions (under any price régime) imply:

(a) more factors at more levels (with a minimum of three levels of each factor);

(b) a systematic arrangement of factor levels or ratios into treatments; and

(c) less emphasis on replication and the sacred cows of arbitrarily chosen significance levels that can bear only a coincidental relation to economic criteria for risky decisions.

Most experimenters readily accept the need for a systematic choice of treatments so as to avoid asking Nature a set of random questions. Agricultural research tradition, however, has overemphasized the importance of replication and, *ipso facto*, has not favoured the consideration of more factors at more levels. But at least in some countries, these cautionary biases towards replication are being overcome (Boyd, 1973).

The traditional analysis of variance approach to ascertain "significant" differences between factor effects implies substantial replication, not merely to increase the accuracy of the analysis of variance, but to measure it as well. For response function estimation, replication is not so essential. Additional levels of a factor can substitute for replications of a particular level since additional observations help to locate the response curve or surface more accurately whether they come from replications or extra factor levels. Indeed, for a given quantity of research resources, extra factor levels help locate the surface better than extra replications. As well, extra levels rather than replications are needed if alternative response models are to be tested.

The essence of the design problem is to obtain as much beneficial information as possible per unit outlay. For a given amount of experimental resources, a balance has to be chosen between plot size, number of factors, number of treatments (i.e. combinations of factor levels) and number of replicates. To give an example, with 27 experimental plots of a given size, some of the many possible choices are as follows:

Replicates	Factors	Levels	Treatments
27	1	1	1
9	1	3	3
3	2	3	9
1	27	1	27
1	1	27	27

Conceptually, as Anderson and Dillon (1968) have shown, optimal choice between such alternatives is feasible. But as a practical matter, heavy reliance has to be made on judgement and intuition. In making such judgements, however, it is crucial that the focus be towards assisting farmers in their decisions. It should not be towards arbitrary levels of significance aimed at satisfying the cautionary motives of the researcher's scientific peer group.

5.2.1. DESIGNS FOR RESPONSE SURFACE ESTIMATION

There are four types of experimental design that are particularly pertinent to the estimation of response surfaces. These designs are the complete factorial, fractional factorial, central composite, and rotatable designs. We will give an example of each. For the reasons outlined in Section 5.2.3 below, while these designs are directly applicable to crop trials involving fertilizers and other husbandry factors, their application to livestock feeding experiments may be a little more complicated except for the case of grazing trials where stocking rate, rather than feed *per se*, is the input variable of interest.

Complete Factorial Designs

The essence of these designs is that we choose our levels of each factor and then take all combinations of factors at all their levels as our treatments. With n levels of k factors, there are therefore n^k-treatments or experimental points in a complete factorial design. Thus for two factors N and P, each taken at three levels, we would have the nine treatments (N, P) shown in Table 5.1. If the factor levels are equally spaced, they can be coded for simplicity of analysis. Thus if the N and P levels of

DIFFICULTIES IN RESPONSE RESEARCH   153

TABLE 5.1

TREATMENTS IN A THREE-LEVEL FACTORIAL
DESIGN INVOLVING TWO FACTORS, N AND P

Level of N	Level of P		
	p_1	p_2	p_3
n_1	$n_1 p_1$	$n_1 p_2$	$n_1 p_3$
n_2	$n_2 p_1$	$n_2 p_2$	$n_2 p_3$
n_3	$n_3 p_1$	$n_3 p_2$	$n_3 p_3$

Table 5.1 each ranged over values of 0, 100 and 200 kilograms per hectare, we could code these factor levels as -1, 0, and $+1$. The nine treatments might then be listed as the experimental points:

$$(-1, -1) \quad (-1, 0) \quad (-1, 1)$$
$$(0, -1) \quad (0, 0) \quad (0, 1)$$
$$(1, -1) \quad (1, 0) \quad (1, 1)$$

Compared to the other response surface designs discussed below, complete factorials can be rather expensive if there are more than two factors, especially if we allow for replication to gain an estimate of experimental error. Relatively speaking, the more limited research resources are, the more wasteful is the complete factorial approach likely to be.

Fractional Factorial Designs

These are obtained by taking only some fraction of the treatments from a complete factorial. For example, from the complete two-factor three-level factorial above, we might use the fractional factorial

$$(-1, -1) \quad \quad (-1, 1)$$
$$ \quad (0, 0) \quad $$
$$(1, -1) \quad \quad (1, 1)$$

or any other fraction that is reasonable and pertinent. Boyd (1973), for example, reports success with a one-sixth fraction of a three-factor six-level factorial in fertilizer trials. Whereas the complete factorial would involve $6^3 = 216$ treatments or plots, the one-sixth fractional involves only 36 treatments or plots per replicate.

Central Composite Designs

These designs are obtained by supplementing the 2^k experimental points of two-level k-factor factorials (or some fraction thereof) by an additional $2k$ experimental points or treatments arranged symmetrically around the factorial, plus a point at the centre of the two-level factorial. With supplementation of a complete factorial, the number of treatments in a central composite design is thus $2^k + 2k + 1$. For example, with two factors a central composite design would have the following coded treatments:

		$[(0, \beta)]$		
	$(-1, 1)$		$(1, 1)$	
$[(-\beta, 0)]$		$[(0, 0)]$		$[(\beta, 0)]$
	$(-1, -1)$		$(1, -1)$	
		$[(0, -\beta)]$		

where the extra treatments augmenting the basic factorial are in square brackets. The factor level $\pm\beta$ is at the choice of the experimenter provided $|\beta| > 1$ and, of course, that the factor levels corresponding to $-\beta$ are not less than zero. Note that design layouts like the above can be interpreted as a "two-dimensional experiment map" with each coded treatment specifying a point in the plane of the two factors, the origin of the map being the central treatment $(0, 0)$. With three factors, we would analogously have a "three-dimensional experiment map" specified by the treatments. As is typical of central composite (and rotatable) designs, the above design does not contain a treatment with both factors at their lowest level. Oftentimes such a treatment would be desired for visual extension purposes. Such might be the case for the treatment with all fertilizers at an actual level of zero in a crop-fertilizer trial, i.e. the treatment $(-\beta, -\beta)$ in the above design if the coding is such that $-\beta$ corresponds to an actual factor level of zero. When such a treatment is desired, it can simply be added to the design.

With only two factors, the central composite design gives no saving in treatment numbers over the complete factorial, but as the number of factors increases, the saving in number of treatments increases, as shown in Table 5.2. More importantly, compared to factorials, the central composite design gives a larger number of levels per factor. For example,

TABLE 5.2
NUMBER OF TREATMENTS PER REPLICATE NEEDED FOR FACTORIAL AND CENTRAL COMPOSITE DESIGNS

Design	2	3	No. of factors 4	5	6
Factorial with three levels	9	27	81	243	729
One-third of three-level factorial		9	27	81	243
Central composite	9	15	25	27[a]	45[a]

[a] Based on augmentation of a fractional (one-half) two-level factorial.

the above composite design contains five levels of each factor compared with only three levels in the two-factor, three-level factorial involving the same number of treatments. This is an important advantage since, as already stressed, the more levels of each factor the better we can define the response surface.

Rotatable Designs

In contrast to the factorial and central composite designs, rotatable designs were developed specifically for response surface estimation in industry (Myers, 1971). Agriculture, of course, differs greatly from industry—we can usually only repeat crop and livestock trials year by year, not day by day; nor can we usually work with a controlled environment. Nonetheless, like the central composite designs (which can be made rotatable by a suitable choice of experimental points—e.g. $\beta = 2^{k/4} = 1\cdot 682$ for $k = 3$), rotatable designs can be relevant for situations involving three or more input factors. Their essential features are: (a) that for given input units of measurement and a polynomial representation or approximation of the response function, they give estimates of response whose variance depends only on the distance from the centre of the design and not on the direction from the centre; and (b) that they enable a necessary estimate of experimental error to be obtained by replication of the central treatment so that it is not essential to replicate the whole experiment, although if the experimenter wishes to, there is no reason why other experimental points in the design should not also be replicated.

An example of a three-factor rotatable design suited to estimating a quadratic or second-order polynomial response function of the form

$$Y = b_0 + b_1 X_1 + b_2 X_2 + b_3 X_3 + b_{11} X_1^2 + b_{22} X_2^2 + b_{33} X_3^2$$
$$+ b_{12} X_1 X_2 + b_{23} X_2 X_3 + b_{13} X_1 X_3 \qquad (5.1)$$

is given in Table 5.3. All told, this design could involve a minimum of sixteen treatments if we allowed but two observations on the central point to give an estimate of experimental error. Generally, however,

TABLE 5.3

CODED TREATMENTS FOR A THREE-FACTOR SECOND-ORDER ROTATABLE DESIGN

	X_1	X_2	X_3
Basic two-level factorial	−1	−1	−1
	1	−1	−1
	−1	1	−1
	1	1	−1
	−1	−1	1
	1	−1	1
	−1	1	1
	1	1	1
Treatments augmented to basic factorial	−1·682	0	0
	1·682	0	0
	0	−1·682	0
	0	1·682	0
	0	0	−1·682
	0	0	1·682
Replicates of central point	0	0	0
	−	−	−
	−	−	−
	0	0	0

some more replications of the treatments would be desirable. Note that though the design of Table 5.3 may involve as few as 16 treatments, each of the three factors appears at five levels. In contrast, 27 treatments would be required for a single replicate of a three-level three-factor complete factorial.

DIFFICULTIES IN RESPONSE RESEARCH 157

Table 5.4 presents an example of the factor levels used in a maize-fertilizer trial based on the design of Table 5.3. Specifications of many other rotatable designs are available in such sources as Cochran and Cox (1957, ch. 8A), Heady and Dillon (1961, ch. 5), Johnson and Leone (1964, ch. 17) and Myers (1971).

TABLE 5.4

LEVELS OF NITROGEN (N), PHOSPHORUS (P) AND POTASSIUM (K) IN A MAIZE-FERTILIZER TRIAL USING A SECOND-ORDER ROTATABLE DESIGN

Coded value:	−1·682	−1	0	1	1·682
Active kg/ha of N:	0	68·93	170·00	271·07	340·00
Active kg/ha of P:	0	48·18	118·80	189·42	237·60
Active kg/ha of K:	0	43·74	108·00	172·26	216·00

5.2.2. CHOICE OF DESIGN

It would be nice to be able to state clearcut rules about choice between alternative multi-level multi-factor designs. But about all that can be said is that, given limited research resources, response surface estimation demands a good spread of observations across the surface rather than exceedingly precise estimation of but one or two points on the surface. "As many important factors as possible in as many treatment combinations as is statistically and economically reasonable", not "Many replicates of a single treatment", should be the guiding rule.

Translated into more practical terms, the above broad principle leads to suggestions along the lines made in part by Anderson (1971) and Boyd (1973):

(a) a priority listing should be made of the potential factors to be studied;

(b) available experimental resources should be assessed to indicate the number of experimental units (plots or animal groups) of different sizes that could reasonably be available for allocation between treatments and replicates with different designs involving alternative numbers of factors moving down the priority listing;

(c) as well as being guided in points (a) and (b) above by any related research, the experimenter should recognize that his proposed experiment(s) might be combined with these other trials for joint analysis—too often trials are run and analysed as single experiments without regard to the fact that, taken in combination with other experiments, they may produce far more information than if organized and analysed in isolation;

(d) in thinking of design possibilities, minimal guidelines to aim for relative to the number (k) of factors involved might be something like:

if $k = 1$, use at least six or seven levels with at least one replication;
if $k = 2$, use at least three-fifths of a 5^2 factorial and if possible a complete 5^2 factorial, with at least one replication;
if $k = 3$, try to have at least five levels of each factor in a fractional factorial with at least one replication; and
if $k \geq 4$, aim to have at least four levels of each factor in a fractional factorial with at least one replication—and if resources are extremely limited, consider a central composite or rotatable design.

In summary, with limited research resources, the problem in choice of design is to achieve an appropriate balance between replication and number of treatments and factors. The more variable the process, the greater the need for replication. Likewise, the larger the number of important factors, the more the need for a greater number of treatments and the less the resources available for replication.

Regardless of the design chosen, the researcher also has to choose where to "centre" it, the distribution of treatment levels, and the factor ranges to be straddled. Here again judgment has to be used. Guided by any relevant prior information, the best that can be done is to guess the region of best operating conditions and make that about the centre of the design. Choice of factor ranges can only be guided by whatever information is available on their biological and potential economic relevance. Traditionally, a symmetrical distribution of factorial treatment levels has been most common. However, as discussed by Anderson and Nelson (1975), it may be more efficient to have a heavier distribution of treatments in the more strongly upward-sloping region of the

DIFFICULTIES IN RESPONSE RESEARCH

response surface so as to better reflect the economic importance of this region.

For some types of response experimentation, particularly crop rotation and livestock grazing trials, there may also be special considerations to be taken into account. With crop rotation trials, as emphasized by Yates (1949) and exemplified by, e.g. Battese and Fuller (1972), Battese, Fuller and Shrader (1972) and Shrader, Fuller and Cady (1966), it is especially desirable to have observations for all phases of the rotations in each year.

In grazing trials, since stocking rate and the matching of pasture demand and supply are such important decision considerations to the farmer, it is important that these be encompassed within the experiment. Too often in the past, stocking rate has not been allowed for as a treatment and attention has been focused only on pasture husbandry factors or, at best, on optimal physical (rather than economic) stocking rate. With these provisos, a fair sampling of the extensive literature is given by Conniffe et al. (1970 and 1972), Cowlishaw (1969), Dillon and Burley (1961), Morley and Spedding (1968), Owen and Ridgman (1968) and Peterson, Lucas and Mott (1965). The work of Jones and Sandland (1974) and Sandland and Jones (1975) is interesting for its conclusion, based on review of an extensive set of studies, that weight gain per animal (g) and stocking rate (s) may reasonably be taken as following the linear relation $g = a - bs$. In consequence, weight gain per hectare (G) relates to stocking rate via the quadratic $G = as - bs^2$. The apparent linear relation between weight gain per animal and stocking rate implies two rates of stocking may be adequate (in combination with other treatment factors) for meat grazing trials. The joint grazing of species (sheep and cattle) is considered by, among others, Bennett et al. (1970), Hamilton and Bath (1970) and Wills and Lloyd (1973).

5.2.3. PEN-FEEDING TRIALS

The factorial and other multi-level designs outlined in Section 5.2.1 above are particularly appropriate for crop-management and stocking rate experiments. In such trials, within a given environment, the experimenter can exercise virtually full control over the treatments studied. But, as discussed by Heady and Dillon (1961, ch. 7), with livestock pen-feeding experiments there is a difficulty in that it may not be possible to control all aspects of the treatments. To give a simple example, we may be able to specify what quantity of feed an animal can have, but not the time span over which it will eat this quantity of feed. Conversely, we may be able to specify the feed period but not the feed quantity to be consumed.

The difficulty shows up in that livestock feed treatments usually specify not symmetrically arranged points but a number of ration lines in the feed input space. Each ration line specifies a particular feed mix. (For example, in a milk-grain pig-feeding trial we might have three ration lines in the milk–grain plane corresponding to: 100 per cent milk; 75 per cent milk, 25 per cent grain; and 50 per cent milk, 50 per cent grain.) As an experiment proceeds, over time the animals "move out along the ration lines". In consequence, for the estimation of pen-fed livestock response functions, the most important element of good design strategy is to ensure that the ration lines give an adequate spread across the biologically feasible region of the response surface. One ingenious way of ensuring this is illustrated by the work of Battese *et al.* (1968) and Holder, Wilson and Williams (1969). In a pig-feeding trial with skim milk and grain, they set the diet treatments as proportions (from 2 to 5 per cent) of animal body weight with the constituent milk and grain allocated across diets on the basis of a two-factor rotatable design.

In conducting pen-feeding experiments, allowance should also be made for sub-*ad libitum* feed levels as discussed by Dean *et al.* (1972), Dent and English (1966) and Duloy and Battese (1967).

5.3 Statistical Estimation†

5.3.1. LEAST-SQUARES REGRESSION

Though response analysis naturally demands the fitting of continuous response functions, it is always a good idea to gain a general picture of the data by first applying an appropriate analysis of variance. Procedures are presented, for example, by Cochran and Cox (1957), Mendenhall (1968), Ostle (1963), Snedecor and Cochrane (1967) and Yates (1954). Then, if there are none of the complications mentioned below, the

† Since statistical analysis is forbidding to the uninitiated, both introductory and more advanced references are listed in what follows.

DIFFICULTIES IN RESPONSE RESEARCH   161

response function may be estimated by means of ordinary least-squares regression. This is a standard procedure for estimating continuous functions and is outlined in most statistical texts such as Draper and Smith (1966), Fox (1968), Huang (1970), Johnson and Leone (1964) and Kmenta (1971, chs. 7–10). Good reviews of the practicalities of regression, its interpretation and pitfalls are provided by Box (1966) and Rao and Miller (1971). Descriptions of the method in terms of response analysis are contained in Heady and Dillon (1961, ch. 4), Bradley and Hunter (1958), Ezekiel and Fox (1959), Myers (1971), Throsby (1961) and Yeh (1964). Draper (1961) discusses the question of missing observations in response surface estimation; Singh and Day (1974) the problem of limited data; and Anderson (1968a), Colwell (1976, ch. 10) and Colwell and Stackhouse (1970) the inevitable problems of assessing raw trial data. Fuller (1969) and Gallant and Fuller (1973) present procedures for spliced or segmented polynomial regressions. Colwell (1976) and Myers (1971, ch. 3), among others, outline the use of orthogonal polynomial regression. The estimation of non-linear regression is discussed by Draper and Smith (1966, ch. 7) and Marquardt (1966 and 1970). The use of the modified regression technique known as ridge regression to ameliorate problems of multicollinearity (i.e. high correlations between the factor data series), as may often occur in livestock feeding trials particularly, is outlined by Brown and Beattie (1975), Hoerl and Kennard (1970a and b) and Marquardt (1975). Ryan (1972) presents an empirical situation well-indicating the difficulties of multicollinearity. Anderson and Nelson (1975) present regression procedures for linear response and plateau models involving intersecting straight lines as discussed in Section 5.5 below.

5.3.2. COMBINING CROSS-SECTION AND TIME-SERIES DATA

Response experiments often involve repeated observations on the same experimental unit; for example, weekly readings of the liveweight of animals in a group, multiple cuttings of hay from a plot, crop rotation sequences on the same field, yields of perennial crops. With such data sets combining cross-section and time-series observations, the error assumptions of ordinary least-squares regression are likely to be upset by autocorrelation due to sequential observations on the same unit not being statistically independent. Discussion of this difficulty and procedures for handling it are presented by Fuller (1976, ch. 5), Fuller and Battese (1973 and 1974), Hall (1975) and Hoel (1964). Application of the procedures is illustrated by Lovell et al. (1974) for livestock and by Battese and Fuller (1972), Battese, Fuller and Shrader (1972), Fuller and Cady (1965) and Shrader, Fuller and Cady (1966) for crop rotation experiments. The standard analysis of variance approach to rotation experiments is presented by Yates (1954) and exemplified by Abraham and Agarwal (1967) and Agarwal (1968).

5.3.3. MULTI-EQUATION AND OTHER MODELS

As illustrated in our broiler and sheep grazing examples of Sections 3.9.2 and 3.9.3, a production process may require a multi-equation model. Empirical examples are provided by the work of Battese et al. (1968) and Townsley (1969) for pigs, Hochman and Lee (1972), Hoepner and Freund (1964) and Kennedy et al. (1976) for broilers, and Dean et al. (1972) for dairy cattle. As these studies indicate and as shown by Duloy and Battese (1967), multi-equation systems are particularly pertinent in livestock processes because the animals exercise choice over time as to the quantity and type of feed they consume. Where such multi-equation models involve jointly determined variables, as discussed by Heady and Dillon (1961, ch. 6) and Dean et al. (1972), simultaneous equations procedures as outlined by such econometric texts as Dhrymes (1970, chs. 4, 5), Huang (1970, chs. 9, 10) and Kmenta (1971, ch. 13) should be used.

Estimation procedures for the analysis of joint or multiple-product response from a single process (such as wool and meat from sheep) are introduced by Gnanadesikan (1963) and Mundlak (1964). A pertinent statistical procedure with multiple products is canonical correlation as outlined by Anderson (1962, chs. 12, 13), Bofinger (1975), Dhrymes (1970, ch. 2), Theil (1971, pp. 317–25) and Vinod (1968). Response oriented applications are presented by Kaminsky (1974) and Reid and Thomas (1973).

Somewhat distinct from the estimation of an input–output function for efficiency analysis is the problem of specifying the underlying physical mechanism of response or of determining the parameters in a known physical law. These questions are discussed by Box (1958), Box and Hill (1967), Box and Hunter (1965), Hill and Hunter (1974), Hunter, Hill and Henson (1969) and Milhorn (1966).

5.3.4. SERIES OF EXPERIMENTS

As noted in Section 5.2.2, efficiency in response research demands that the results from different experiments (both across space and time) should be analysed as a whole whenever feasible. The statistical aspects (largely relating to non-homogeneous error variance between experiments) and/or empirical examples of such combined analyses are presented by Boyd (1973), Boyd et al. (1970), Boyd, Tong Kwong Yuen and Needham (1976), Cochran and Cox (1957), Colwell (1967–8), Laird et al. (1969), Laird and Cady (1969), Pesek (1973), Ryan (1972), Ryan and Perrin (1973) and Tollini and Seagraves (1970).

Efficient response research also requires that, so far as possible, whatever prior information is available be taken into account in estimating the response function. Such information may relate to the general form of the function or the size of particular parameters. This implies the use of Bayesian procedures as outlined by Blight and Ott (1975), Chowdhury Nagadevara and Heady (1975), Halter and Dean (1971) and Zellner (1971).

Johnson (1971) notes the biases that may occur in aggregating technical unit response function estimates to whole-farm level.

5.3.5. ECONOMIC VERSUS STATISTICAL SIGNIFICANCE

So far in our discussion of statistical estimation we have been talking as if there was just one function—the response function—to be estimated. However, because of the pervasiveness of uncertainty in response due to uncontrolled factors, this is far too simple a view. As was shown in Chapter 4, what we would *ideally* like to have is an estimate of the probability distribution of profit conditional on the decision variables. This implies a different probability distribution of yield for each possible combination of the decision variables, i.e. not one function but a large—if not infinite—number. Choice of best operating conditions then corresponds to choice of the probability distribution giving greatest expected utility. That's the decision problem from the farmer's view. In contrast, regrettably if not shamefully, traditional statistical methods of appraising the worthwhileness of response function estimates have emphasized a quite unrelated decision problem.

Via tests of statistical significance (the "cult of the asterisk") involving mechanical application of arbitrary probabilities of accepting a false hypothesis, traditional procedures—as discussed by Schlaifer (1959)—have aimed at protecting the researcher from "scientific error". In doing so, these procedures have led to a far greater error of research-resource waste. The farmer's problem is not whether or not there is a 5 per cent or less chance that a crop-fertilizer response function exists. His problem is how much fertilizer to use. Even if the estimated function is only statistically significant at the 50 per cent level of probability, it may still be exceedingly profitable in expected utility terms for the farmer to base his decisions on the estimated function.

As Dillon and Officer (1971) note, by emphasizing the Type I error of accepting a false hypothesis at arbitrary significance levels while ignoring the Type II error of rejecting a true hypothesis, traditional significance test procedures: (i) are heavily biased to the *status quo*; and (ii) do not take into account the respective costs and benefits of different decisions based on the statistical estimates being tested. In the mechanical fashion in which they are usually applied, significance tests have no economic basis and hence no relevance, except by chance, to farmer decisions about best operating conditions. Indeed, as noted by Anderson (1971), Boyd (1973) and Yates (1964), significance tests on response estimates are often posed in a ridiculous form—after thousands of feed and fertilizer experiments it is quite unreasonable to advance the null hypothesis that yield response in livestock or crops is unaffected by feed or fertilizer respectively.

What then are we to do in appraising statistical estimates of response? On the one hand we have the practical but irrelevant procedures of significance testing, and on the other, the relevant but largely impractical procedures of full-scale expected utility analysis. Unsatisfactory though it be to those seeking the impossible goal of scientific objectivity, the only reasonable procedure is the common sense one of judging response estimates: (a) by their reasonableness in terms of the researcher's subjective judgment relative to the strength of the evidence and whatever prior information is available; and (b) by intuition or by more formal appraisal of the expected benefits of alternative decisions with and without the estimates—as initially suggested by Havlicek and Seagraves (1962) and as elaborated in more formal decision theoretic terms by Anderson, Dillon and Hardaker (1976), Hadley (1967, pp. 492-6), Lindley (1965, Section 5.6) and Winkler (1972).

If significance tests and their associated confidence limits must be calculated, logic demands that this should be done not just for the response function *per se* but for the profit-related quantities derived from the response function. Such procedures are presented by Fuller (1962), Hoffnar (1963) and Doll, Jebe and Munson (1960). Better still, the variance of profit may be assessed—as exemplified by Battese and Fuller (1972) and Ryan and Perrin (1973)—thereby allowing a large measure of economic choice between alternative risks rather than mere reliance on irrelevant levels of significance.

Finally, as stressed by Anderson (1971) and Raiffa and Schlaifer (1961), in reporting the statistical results of response analysis it is important that at least the sufficient statistics be presented. These are those statistics which summarize all the information from a sample so that any additional statistics are uninformative. For ordinary regression analysis these will consist of the estimated regression coefficients, the associated variance-covariance matrix, the error mean square and the sample size. Best of all is a complete tabulation of the data. Such reporting is important so that others can make their own judgements about the analyses presented, and also to facilitate joint analysis with other data sets.

5.4 Response Variability over Space and Time

For any particular response process, marked variations in yield occur as the process is repeated across time and space. As discussed in Chapter 4 in relation to risk, the reason for such variation is that even though identical levels of the decision variables $X_1, \ldots, X_n$ be used, variations across space and time in the predetermined variables $X_{n+1}, \ldots, X_k$ and the uncertain variables $X_{k+1}, \ldots, X_m$ cause yields to vary. Such variation is especially frequent in crop and grazing processes due both to variation in the predetermined fertility status of the soil and to variation in the uncontrolled elements of the climatic régime under which the process operates.

Our emphasis in Chapter 4 was on the implications of uncertain response variability for the appraisal of best operating conditions. In this section our concern is the research implications of response variability.

Response variation due to the influence of either predetermined or uncertain factors is only economically relevant if there is interaction between these factors and the decision variables. From all the evidence available, such interaction must be invariably expected. It is therefore important that response research be designed so far as is economically reasonable to bring all the uncontrolled factors—both predetermined and uncertain—to account along with the decision variables.

From a decision making view—and *ipso facto* from a research view—the predetermined and uncertain factors are qualitatively quite distinct. The uncertain uncontrolled factors cause uncertainty; the predetermined uncontrolled factors, along with the decision variables, condition the uncertainty. By their nature, the influence of the uncertain factors can only be handled in a probabilistic framework. But, so long as the predetermined factors are measured, there need be no uncertainty due to them *per se*. Accordingly, response research should incorporate variation in the predetermined factors so as to enable appraisal of their influence as, for example, outlined in Section 4.5.3 in relation to equations (4.90), (4.91) and (4.92) for expected yield, variance and skewness respectively. In terms of experimentation and analysis, this means that ideally the design of experiments and their subsequent analysis should include, so far as possible, the relevant predetermined factors $X_{n+1}, \ldots, X_k$ as experimental factors. If this cannot be done, measurements should at least be taken on these factors for each experimental treatment and these data incorporated in subsequent analysis.

The inclusion of predetermined factors in the (expected) yield response function has been best developed for crop-fertilizer studies. In this context the response function is often referred to as a GENERALIZED YIELD EQUATION—see, for example, the work of Anon. (1974a, pp. 149–54), de Oliveira (1973), Jensen and Pesek (1959), Pesek (1973), Ryan (1972) and Voss and Pesek (1962a). The generalized (expected) yield equation may be of any desired appropriate algebraic form. Ryan and Perrin

(1974) illustrate a quadratic polynomial; de Janvry (1972a) a generalized power function which provides scope for more direct consideration of the biological aspects of response.

Examples of the predetermined factors that have frequently been included in crop response are soil phosphorus, soil nitrogen or organic matter, soil potassium, soil pH, depth of soil, clay content and—via dummy variables—location, soil type and trace element status. As discussed by Ryan (1972), two approaches may be taken to the inclusion of nutrient soil test values. One is to convert initial soil nutrients to equivalent applied nutrient units and then aggregate the two sources of nutrient in the analysis. Discussion and examples are provided by, among others, Brown, Jackson and Petersen (1962), Colwell (1967-8 and 1976), Valdés (1967) and Voss and Pesek (1962b). The other approach, far simpler, is to include the soil test variables directly as independent variables in the regression analysis of response, as exemplified by de Janvry (1972a), Rao and Sengupta (1965), Russell (1968b) and Ryan and Perrin (1973). See also Inkson (1964).

The major uncertain factors in response are climatic variables, though pest and disease effects may often also be important. By their nature, the uncertain uncontrolled factors have to be brought to account through the probability distribution of yield. As noted in Section 4.5.3, this may be carried out by estimating empirical relationships between the relevant parameters of the yield distribution and the levels of the decision and predetermined variables. With this gross approach it is not essential to have recorded observations on the uncertain variables themselves. Such observations, however, will certainly be a most useful complement to analysis. Alternative to the gross approach, full analytical specification of the probability distribution of yield as the joint distribution of the uncertain variables conditional on the decision and predetermined variables may be attempted. To do so, however, it is essential that observations also be available on the uncertain variables—at least to the extent they can be specified. Because full specification is usually impossible, this approach will typically lead to underestimation of the risks being faced. Whichever approach is used—gross or analytical—it is necessary to run the experiment across both time and space so as to allow variation in climate and other uncertain factors to operate. A one-time one-location experiment will provide little information for risk analysis.

As reviewed by Ryan (1972), three approaches have been taken to the measurement of climatic factors. Bondavalli, Colyer and Kroth (1970), Byerlee and Anderson (1969), Englestad and Doll (1961), R. J. Hildreth (1957), Runge (1968), Ryan and Perrin (1973) and Voss and Pesek (1967), for example, use simple measures of temperature and/or rainfall as independent variables in the response function; Mandac (1974) and Montaño and Barker (1970) use solar radiation. Others such as Ewalt, Doll and Decker (1961), Havlicek and Seagraves (1962), Knetsch (1959), Nix and Fitzpatrick (1969) and Parks and Knetsch (1959) have used composite climatic indices. And Colwell and Esdaile (1968), Fitts et al. (1959) and Pesek (1973) report the use of an orthogonal polynomial trend procedure developed by Fisher (1924) whereby an attempt is made to capture the time-distribution aspects of such variables as rainfall and temperature.

Closely related to the appraisal of climatic effects is the analysis of crop response under irrigation. In this context, full appraisal cannot avoid taking account of uncertain intra-seasonal climatic sequences if the optimal time-pattern of irrigation is to be chosen. Some relevant studies are presented by Downey (1972), Dudley and Burt (1973), Hogg et al. (1969), Minhas, Parikh and Srinivasan (1974), Stewart et al. (1974), Stewart and Hagan (1973), Wu and Liang (1972), Yaron et al. (1972 and 1973) and Yaron and Bresler (1970).

As already noted, pests and diseases may sometimes be important uncertain factors in crop and livestock response. Little analysis of their effects in a response context appears to be available except for the Philippine rice work of I.R.R.I. (1969 and 1972), Mandac (1974) and Roumasset (1974). Relatedly, Roumasset (1976, ch. 7) discusses the risk-reducing role of chemical control factors for pests and diseases.

Some comment on the question of how response experiments might best be located over space so as to appraise variability is given by Anderson (1967b and 1973), Hartley (1964) and Hoffnar and Johnson (1966). The general question of risky response experimentation and the possibilities of analysis as determined by the data available are best summarized by the extension of the work of Anderson (1973) presented by Anderson, Dillon and Hardaker (1976). It is also relevant to note the development of simulation modelling approaches to response and its variation as covered, for example, by Anderson (1974c), Blackie and

Schneeberger (1971), Dent and Anderson (1971), Mihram (1972), Nix and Fitzpatrick (1969), Reid and Thomas (1973) and Windsor and Chow (1971).

Reinforced by a lack of appreciation of the mathematical basis of response surface analysis, the problem of response variability over space and time has frequently been used by agricultural scientists to argue against the response function approach to crop and livestock response. The argument is quite invalid, however, even without consideration of the techniques for handling response variability mentioned above. Temporal and spatial variability is just as much a problem—if not more so—in the traditional qualitative approach of looking for significant differences in response between some few levels of a factor. Certainly, in the commercial arena, the burgeoning demand by farmers and extension agencies for sophisticated recommendations on feed and fertilizer use based on response surface analyses attests to the inadequacies of the traditional qualitative approach to crop and livestock response. More and more, the farmer's question is not "Should I use this or that?" but "How much?"

5.5 Choice of Response Model

The comparative assessment of alternative algebraic forms for the (expected) response function has been a topic of abiding empirical interest. Major emphasis has been on the comparison of quadratic polynomials in X_i and $X_i^{.5}$ and variants of the power, Mitscherlich and hyperbolic functions. Typical examples are provided by the work of Abraham and Rao (1966), Anderson (1957), Anderson and Nelson (1971), Cady and Laird (1969), Colwell (1976, ch. 3), Heady and Dillon (1961, chs. 8–15), Johnson (1953) and Jonsson (1974). Stemberger (1957) has compared quadratics and the form-free model proposed by C. G. Hildreth (1957). The usual criteria applied in such comparisons—see Heady and Dillon (1961, chs. 3, 6)—have been an amalgam of (a) statistical measures of goodness of fit and "significance", (b) *a priori* considerations relating to the biology and economics of the response process, (c) subjective judgement, and (d) computational ease. The general conclusion has been in favour of the quadratic in X_i or $X_i^{.5}$ as against other forms, with some preference for the square-root quadratic attributable to its non-symmetrical and flatter shape in X_i space. These conclusions are not unexpected since the true response function is always unknown and the polynomial form can be justified as a Taylor series approximation of the unknown function, as shown by Heady and Dillon (1961, ch. 6).

Appraisals of the above type, however, have been largely misdirected. As argued by Anon. (1974a) and Perrin (1976) in extending the seminal ideas of Anderson (1968a) and Havlicek and Seagraves (1962) on response model evaluation, the criteria for choosing between alternative models or theories of response analysis should relate to the value of the information they provide. If two models have the same costs in terms of data requirements and application, the preferred model should be the one which provides farmers (ideally society) with the greatest expected utility. If the models differ in their costs, this difference should also be allowed for in the utility appraisal. Of course, utility appraisal is an ideal unlikely of achievement. Expected profit provides a workable and reasonable surrogate giving, for risk averters, an upper bound on the value of information.

5.5.1. LINEAR RESPONSE AND PLATEAU MODEL

The decision theoretic or value of information approach to model choice is well illustrated by Perrin (1976) in his comparison, using Brazilian data, of a generalized quadratic model and the linear response and plateau (LRP) model as postulated for crop-fertilizer response by Waugh, Cate and Nelson (1973). Ignoring their use of relative yields which are an economic cul de sac, the LRP model is based on the law of the minimum of Liebig (1855) who argued: "The crops on a field diminish or increase in exact proportion to the diminution or increase of the mineral substances conveyed to it in manure . . . by the deficiency or absence of one necessary constituent, all the others being present, the soil is rendered barren for all those crops to the life of which that one constituent is indispensable." Blackman (1905) restated Liebig's law as a principle of limiting factors and Swanson (1963) has given a modern statement as "Production is assumed to increase at a constant rate with respect to each factor until some other factor is limiting".

For each decision variable X_i ($i = 1, \ldots, n$), the LRP model (in its simplest form) implies a two-part response function of the linear form

$$Y = \begin{cases} a + bX_i & \text{for } 0 \leq X_i \leq X_i^* \quad (5.2a) \\ a + bX_i^* & \text{for } X_i \geq X_i^* \quad (5.2b) \end{cases}$$

where a and b are non-negative. The parameter a is referred to as the threshold yield for X_i and the parameter X_i^* as the critical level of X_i. The value of X_i^* will vary according as other factors are limiting. If no other factors are in limiting supply to the crop, X_i^*—and hence also Y—will take on its maximum value as determined by the crop's genetic constitution and climatic location.

Response function (5.2) implies linear positive response (i.e. $MP_i = b$) to X_i until the critical level of input, X_i^*, is reached; for inputs of X_i beyond X_i^*, yield remains constant (i.e. $MP_i = 0$) at the plateau level $(a + bX_i^*)$. Hence the name "linear response and plateau model" given to the response function. Except in so far as they influence the critical level X_i^*, the model assumes no interaction between X_i and other decision or uncontrolled factors.

As explained and illustrated by Anderson and Nelson (1975), Anon. (1974a), Cate (1976), Waugh, Cate and Nelson (1973) and Perrin (1976), the LRP model is very easily applied on a site by site basis. In terms of ease of field application and integration with soil type and soil test information, it has significant operational advantages over the use of a (perhaps better fitting) multi-factor quadratic polynomial or other curvilinear form of response model. The above authors also suggest the LRP model leads to simpler economic analysis since its two-part linear form implies X_i should either be used at the level X_i^* if $p_i/p_y < b$ or otherwise not at all. Such appraisal takes no account of risk in X_i^* due to the influence of the uncontrolled uncertain factors. Full risk analysis would involve utility analysis (as per Chapter 4) using the LRP model of expected yield. Suffice to note that for a risk-averse decision maker and increasing marginal risk, the optimal level of X_i would be less than the expected critical level $E(X_i^*)$.

That the simple LRP model of equation (5.2) (or a slightly more complicated version involving a series of intersecting straight lines) may often serve to depict response adequately in statistical terms is well attested by a variety of evidence. That of Anon. (1974a) and Waugh,

Cate and Nelson (1973) using South and North American data has already been noted. Anderson and Nelson (1975) and Bartholomew (1972) found the model satisfactory for nitrogen response in a number of countries. Boyd (1972), Boyd *et al.* (1970) and Boyd, Tong Kwong Yuen and Needham (1976), in an extensive series of analyses of U.K. fertilizer trials with cereals, sugar-beet and potatoes found strong support for a two-part linear model analogous to the Liebig model. For livestock, Fawcett (1973) argues the relevance of a Liebig-type model for pig-feeding response. Upton and Dalton (1976) adduce further evidence.

The important question, however, is not whether any particular type of model is statistically or biologically better than others, but whether it can serve farmers (or society) better. To answer this question for his comparison of the LRP and generalized quadratic function using Brazilian data, Perrin (1976) used a set of 54 experiments conducted at different sites in the State of Minas Gerais over a period of three years. The two response models were estimated from a random sample of 26 of the experiments. The predicted best operating conditions of the two models were then evaluated on a site-by-site basis in terms of the *ex post* profit results they implied on the basis of the data from the remaining 28 experiments. Preferably, this simulation might have been repeated a number of times with different samplings of experiments in the estimation and test groups. The results of the single simulation showed the quadratic model to be rather better than the LRP model if no soil test data were used. With soil test data taken into account, the LRP model was marginally better than the quadratic. No generality attaches to these results they relate only to a particular region of Brazil (and strictly only to the climatic period covered). However, the results do indicate the feasibility of relevant economic comparisons between models. They also show the LRP model to be worthy of further consideration. Analogous appraisal is needed for other regions since, in terms of field application, the LRP model is substantially easier to apply than the more usual continuous curvilinear multi-factor model.

5.6 Economics of Response Research

The criterion for deciding how much to invest in response research is the same as for any other investment: invest until the marginal opportunity cost is equal to the marginal expected benefit or, same thing, until expected utility is maximized. In the response research context this implies investing until the marginal opportunity cost of further research (and complementary extension) is equal to the expected value of the information it will generate. More so than for non-research investments, which can be difficult enough to appraise—see Anderson, Dillon and Hardaker (1976, ch. 8), this principle is difficult to apply formally to choice between alternative levels of research investment. These decisions must inevitably rely heavily on intuition and subjective judgement. Such choices, however, can certainly be aided by a knowledge of the decision-theoretic principles related to the purchase of further information as enunciated by, for example, Lin (1974), Pratt, Raiffa and Schlaifer (1965, chs. 23, 24), Raiffa and Schlaifer (1961, ch. 5) and Schlaifer (1959). More general consideration of the economics of research and information is given by Dillon (1975), Evenson and Kislev (1975) and Hirshleifer (1971).

At the more practical level, a few people have discussed the economics of response research *per se*. In historical sequence, the major contributions are probably those of Yates (1952), Finney (1960), Havlicek and Seagraves (1962), Anderson (1968a), Anderson and Dillon (1968 and 1970), Seagraves (1970), Ryan and Perrin (1973 and 1974) and Perrin (1976).

Anderson and Dillon (1968 and 1970) give the most direct consideration of the response research investment problem. To avoid the complications of macro cost and benefit aspects of publicly sponsored research, their analysis is confined to *ex ante* appraisal of how much an individual producer should invest in his own response research. In doing so, they take account of existing prior knowledge, the farmer's attitude to risk, the life of the knowledge gained, the size of the universe to which the research applies and the cost of the research. The criterion used is maximization of the producer's expected utility specified in terms of profit mean and variance. Their empirical example illustrates the use of simulation procedures in the *ex ante* evaluation of research alternatives.

At the macro level, Ryan and Perrin (1973 and 1974) have estimated the potential short-run benefits to producers and the long-run benefits to consumers of fertilizer response research on potatoes in Peru. Their work also gives an evaluation of the value of soil classification and testing in Peru. This topic is well treated at the micro level of the individual producer by Anon. (1974a) and Perrin (1976) using the value of information approach. Related but less complete contributions are those of Bie and Ulph (1972) for soil classification and Colwell (1968) for soil testing.

5.7 Farm Versus Experimental Response

A problem that has received insufficient attention in crop and livestock response research is the discrepancy that typically occurs between experimental and farm response. In general, response under experimental conditions significantly exceeds the response achieved under workaday farm conditions. For this reason response functions based on farm trials as discussed by Jones and Hocknell (1962), Herdt and Bernsten (1975) and Herdt, de Datta and Neeley (1975), or farm survey data of the type discussed by Yates and Boyd (1965), Hoffnar and Johnson (1966), Anderson (1967b) and Davidson, Martin and Mauldon (1967) could be well worthwhile.

As Davidson and Martin (1965 a and b and 1967) have shown in an extensive analysis of Australian crop and livestock data, farm yields tend to approach experimental yields the smaller the scale of farming and the more intensive the use of labour in production. In other words, the closer the labour and capital intensity of farming is to the intensity of labour and capital typical in experiments, the less the discrepancy between farm and experimental yields. Some of Davidson and Martin's data testifying to this in crop production is shown in Table 5.5. Representing farm yields as a percentage of experimental yields by W and the average area of the crop per farm by A, the data of Table 5.5 is well described by the regression estimate:

$$W = 102 \cdot 12 - 15 \cdot 75 \log A \quad (R^2 = 0 \cdot 81). \qquad (5.3)$$

Obviously, until further such studies are available and quantitative relations established between farm and experimental yields so that suitable adjustment can be made both in experimental design and analysis, commercial recommendations based on experimental response functions must be less than adequate. Indeed, as Candler (1962) and Dent (1966) have suggested, such unadjusted recommendations can only be regarded

TABLE 5.5

RELATION BETWEEN FARM AND EXPERIMENTAL YIELDS
(AUSTRALIAN DATA)

Location and type of crop	Number of observations	Average area of crop on farms (acres)	Mean farm yield/ mean experimental yield (per cent)
Carnarvon beans	2	5	93
Victorian tobacco	2	5	95
Queensland sugar-cane	51	60	76
Murrumbidgee rice	12	80	65
Victorian wheat	35	144	57
Avondale wheat	7	151	64
Dundas wheat	8	302	65
Chapman wheat	22	485	72
Narambeen wheat	10	600	61
Merredin wheat	29	700	57

Source: Davidson and Martin (1967).

as untested hypotheses that have every chance of being inadequate. Alternatively, recommendations should be based on response functions estimated under farm conditions.

5.8 Making Farm Recommendations

It is a long step from much of the theory we have exposited to the real world of the farmer. This is not to say the principles we have outlined are useless. Appropriately taken into account, they give strong guidance to the formulation of recommendations. But in themselves, these principles cannot directly give precise recommendations—the lack of data and the messiness of the world are too great. Certainly all the data needed for full application of crop and livestock response analysis in all its potential detail will never be available, not least because it would not be worth its cost. Just as surely much more data than is currently available will become available, as will short-cut procedures and graphic or computerized aids to the individual farmer's choice of input levels. Some such aids are already available, as exemplified by Anon. (1974a and b), Campbell and Keay (1970) and Waugh, Cate and Nelson (1973)—not to mention Heady's famous pork costulator (Heady and Dillon, 1961, p. 301).

Generally, however, it will not be feasible for every farmer to receive individual guidance on best operating conditions. Recommendations will have to be made on a regional or group basis. In doing so the following points, some of which are elaborated by Anderson (1976) and Anderson *et al.* (1976), should be borne in mind:

(a) The group or region to which the recommendation is meant to apply should not be vague.
(b) It should be recognized that most farmers operate under opportunity cost constraints imposed by other opportunities and the need to borrow funds.
(c) Because farmers differ in their risk preferences and probability judgements, and in their managerial style, no particular recommendation is likely to be best for all farmers in a region or group.
(d) So far as possible, not just one but a few alternative recommendations and information on their possible range of outcomes due to yield and price risk should be given.
(e) In basing recommendations on experimental data, allowance should be made for the difference expected between farm and experimental response, and it should not be forgotten that traditional tests of statistical significance have little relevance to the farmer's decision problem.

References

The following list of references, all of which are cited in the text, is quite extensive but far from complete. In particular, it tends to emphasize recent rather than early work and barely touches the non-English literature. Complementary listings are to be found in Heady and Dillon (1961), Natl. Res. Council (1961) and Ryan (1972).

ABRAHAM, T. P. (1965) Optimal fertilizer dressings and economics of manuring. *Indian Jour. Agric. Econ.* **20**, 1–20.
ABRAHAM, T. P. and AGARWAL, K. N. (1967) Yield, effect on soil fertility and economics of crop rotation with and without groundnut. *Indian Jour. Agric. Sci.* **37**, 560–71.
ABRAHAM, T. P. and RAO, V. Y. (1960) Economic analysis of fertilizer trial data on paddy and wheat. *Indian Jour. Agric. Sci.* **30**, 1–17.
ABRAHAM, T. P. and RAO, V. Y. (1966) An investigation of functional models for fertilizer response studies. *Jour. Indian Soc. Agric. Stats.* **18**, 45–61.
AGARWAL, K. N. (1968) Analysis of experiments on crop rotation. *Jour. Indian Soc. Agric. Stats.* **20**, 26–43.
ALCANTARA, R. and PRATO, A. A. (1973) Returns to scale and input elasticities for sugarcane: The case of São Paulo, Brazil. *Amer. Jour. Agric. Econ.* **55**, 577–83.
ALLEN, R. G. D. (1938) *Mathematical Analysis for Economists*. Macmillan, London.
ANDERSON, J. R. (1967a) Economic interpretation of fertilizer response data. *Rev. Mktng. Agric. Econ.* **35**, 43–57.
ANDERSON, J. R. (1967b) Some practical notions for estimating agricultural production functions. *Jour. Australian Instit. Agric. Sci.* **33**, 194–5.
ANDERSON, J. R. (1968a) A note on some difficulties in response analysis. *Australian Jour. Agric. Econ.* **12**, 46–53.
ANDERSON, J. R. (1968b) Economics and animal production research. *Proc. Australian Soc. Animal Prodn.* **7**, 428–33.
ANDERSON, J. R. (1971) Guidelines for applied agricultural research: Designing, reporting and interpreting experiments. *Rev. Mktng. Agric. Econ.* **39**, 3–14.
ANDERSON, J. R. (1973) Sparse data, climatic variability, and yield uncertainty in response analysis. *Amer. Jour. Agric. Econ.* **55**, 77–82.
ANDERSON, J. R. (1974a) Sparse data, estimational reliability and risk-efficient decisions. *Amer. Jour. Agric. Econ.* **56**, 564–72.
ANDERSON, J. R. (1974b) Risk efficiency in the interpretation of agricultural production research. *Rev. Mktng. Agric. Econ.* **42**, 131–84.
ANDERSON, J. R. (1974c) Simulation: Methodology and application in agricultural economics. *Rev. Mktng. Agric. Econ.* **42**, 3–55.
ANDERSON, J. R. (1975) One more or less cheer for optimality. *Jour. Australian Instit. Agric. Sci.* **41**, 195–7.
ANDERSON, J. R. (1976) On formulating advice to farmers from agronomic experiments. *Univ. of New England Dept. Agric. Econ. and Bus. Mgmt. Misc. Publicn.* 3, Armidale.

ANDERSON, J. R. *et al.* (1976) *From Agronomic Data to Farmer Recommendations: An Economics Training Manual.* CIMMYT, Mexico City.

ANDERSON, J. R. and DILLON, J. L. (1968) Economic considerations in response research. *Amer. Jour. Agric. Econ.* **50**, 130–42.

ANDERSON, J. R. and DILLON, J. L. (1969) A comparison of response surface and factorial designs in agricultural research: Comment. *Rev. Mktng. Agric. Econ.* **37**, 130–2.

ANDERSON, J. R. and DILLON, J. L. (1970) Economic considerations in response research: Further comment. *Amer. Jour. Agric. Econ.* **52**, 609–10.

ANDERSON, J. R., DILLON, J. L. and HARDAKER, J. B. (1976) *Agricultural Decision Analysis*. Iowa State Univ. Press, Ames.

ANDERSON, R. L. (1957) Some statistical problems in the analysis of fertilizer response data. In E. L. Baum *et al.* (Eds.), *Economic and Technical Analysis of Fertilizer Innovations and Resource Use*, Iowa State Univ. Press, Ames, pp. 187–206.

ANDERSON, R. L. and NELSON, L. A. (1971) Some problems in the estimation of single nutrient response functions. *Bul. Intl. Statist. Instit.* **44**, Part 1, 203–22.

ANDERSON, R. L. and NELSON, L. A. (1975) A family of models involving intersecting straight lines and concomitant experimental designs useful in evaluating response to fertilizer nutrients. *Biometrics* **31**, 303–18.

ANDERSON, T. W. (1962) *An Introduction to Multivariate Statistical Analysis*. Wiley, New York.

ANON. (1974a) *Agronomic-Economic Research on Tropical Soils: Annual Report for 1973*. Soil Sci. Dept., North Carolina State Univ. (Contract AID/csd 2806), Raleigh.

ANON. (1974b) Deciding how much fertilizer to use. *Rural Research in C.S.I.R.O.* No. 86, 4–13.

ARCUS, P. L. (1963) An introduction to the use of simulation in the study of grazing management problems. *Proc. N.Z. Soc. Animal Prodn.* **23**, 159–68.

ARROW, K. J. *et al.* (1961) Capital-labor substitution and economic efficiency. *Rev. Econ. Stats.* **43**, 225–50.

BAIRD, B. L. and MASON, D. D. (1959) Multivariable equations describing fertility-corn yield response surfaces and their agronomic and economic interpretation. *Agronomy Jour.* **51**, 152–6.

BALAAM, L. N. (1975) Response surface designs. In V. J. Bofinger and J. L. Wheeler (Eds.), *Developments in Field Experiment Design and Analysis*, Cwlth. Agric. Bur., Slough, pp. 11–32.

BARROW, N. J. (1973) Relationship between a soil's ability to adsorb phosphate and the residual effectiveness of superphosphate. *Australian Jour. Soil Res.* **11**, 57–63.

BARROW, N. J. and CAMPBELL, N. A. (1972) Methods of measuring the residual value of fertilizers. *Australian Jour. Expt. Agric. Animal Husb.* **12**, 502–10.

BARTHOLOMEW, W. V. (1972) Soil nitrogen: Supply processes and crop requirements *North Carolina State Univ. Soil Fertility Evaluation and Improvement Project Tech. Bul.* 6, Raleigh.

BATTESE, G. E. *et al.* (1968) The determination of optimal rations for pigs fed separated milk and grain. *Jour. Agric. Econ.* **19**, 355–64.

BATTESE, G. E. and FULLER, W. A. (1972) Determination of economic optima from crop-rotation experiments. *Biometrics* **28**, 781–92.

BATTESE, G. E., FULLER, W. A. and SHRADER, W. D. (1972) Analysis of crop-rotation experiments, with application to the Iowa Carrington-Clyde rotation-fertility experiments. *Iowa State Univ. Agric. Expt. Sta. Res. Bul.* 574, Ames.

BAUM, E. L. *et al.* (Eds.) (1956) *Methodological Procedures in the Economic Analysis of Fertilizer Use Data.* Iowa State Univ. Press, Ames.
BAUM, E. L. *et al.* (Eds.) (1957) *Economic and Technical Analysis of Fertilizer Innovations and Resource Use.* Iowa State Univ. Press, Ames.
BENNETT, D. *et al.* (1970) The effect of grazing cattle and sheep together. *Australian Jour. Expt. Agric. Animal Husb.* **10**, 694–709.
BERINGER, C. (1961) An economic model for determining the production function for water in California. Calif. Agric. Expt. Sta. Giannini Foundation, Mimeo Rept. 240, Berkeley.
BICKING, C. A. and GILLESPIE, R. H. (1963) Exploring with experimental design. *Industrial Quality Control* **19** (12), 17–21.
BIE, S. W. and ULPH, A. (1972) The economic value of soil survey information. *Jour. Agric. Econ.* **23**, 285–95.
BILAS, R. A. (1967) *Microeconomic Theory: A Graphical Analysis.* McGraw-Hill, New York.
BLACKIE, M. J. and SCHNEEBERGER, K. C. (1971) Simulation of the growth response of dryland and irrigated corn. *Canadian Jour. Agric. Econ.* **19**, 108–12.
BLACKMAN, F. F. (1905) Optima and limiting factors. *Ann. Botany* **19**, 281–95.
BLANDFORD, D. and CURRIE, M. (1975) Price uncertainty – the case for government intervention. *Jour. Agric. Econ.* **26**, 37–51.
BLAXTER, K. L. (1961) Economics and animal husbandry. *Jour. Agric. Econ.* **14**, 308–13.
BLAXTER, K. L. (1962a) Technical experimentation as a source of input–output data. In O.E.C.D., *Inter-disciplinary Co-operation in Technical and Economic Agricultural Research* (O.E.C.D. Documentation in Food and Agriculture No. 50), Paris, pp. 93–5.
BLAXTER, K. L. (1962b) *The Energy Metabolism of Ruminants.* Thomas, Sprinfield.
BLAXTER, K. L. (1964) The efficiency of feed conversion by livestock. *Jour. Roy. Agric. Soc. England* **125**, 87–99.
BLAXTER, K. L., GRAHAM, N.McC. and WAINMAN, F. W. (1956) Some observations on the digestibility of food by sheep, and on related problems. *Brit. Jour. Nutr.* **10**, 69–91.
BLIGHT, B. J. N. and OTT, L. (1975) A Bayesian approach to model inadequacy for polynomial regression. *Biometrika* **62**, 79–88.
BOFINGER, V. J. (1975) An introduction to some multivariate techniques with applications in field experiments. In V. J. Bofinger and J. L. Wheeler (Eds.), *Developments in Field Experiment Design and Analysis*, Cwlth. Agric. Bur., Slough, pp. 73–84.
BOFINGER, V. J. and WHEELER, J. L. (Eds.) (1975) *Developments in Field Experiment Design and Analysis.* Cwlth. Agric. Bur., Slough.
BONDAVALLI, B., COLYER, D. and KROTH, E. M. (1970) Effects of weather, nitrogen and population on corn yield response. *Agronomy Jour.* **62**, 669–72.
BOULDING, K. E. (1958) *Economic Analysis.* Harper, New York.
BOX, G. E. P. (1954) Exploration and exploitation of response surfaces: Some general considerations and examples. *Biometrics* **10**, 16–60.
BOX, G. E. P. (1958) Use of statistical methods in the elucidation of basic mechanisms. *Bul. Intl. Stat. Instit.* **36**, 215–25.
BOX, G. E. P. (1966) Use and abuse of regression. *Technometrics* **8**, 625–9.
BOX, G. E. P. and HILL, W. J. (1967) Discrimination among mechanistic models. *Technometrics* **9**, 57–71.
BOX, G. E. P. and HUNTER, W. G. (1965) The experimental study of physical mechanisms. *Technometrics* **7**, 23–42.

Box, M. J. and Draper, N. R. (1971) Factorial Designs, the $|X'X|$ criterion, and some related matters. *Technometrics* **13**, 731–42. Corrigendum **15**, 430.

Box, M. J. and Draper, N. R. (1974) On minimum-point second-order designs. *Technometrics* **16**, 613–16.

Boyd, D. A. (1972) Some recent ideas on fertilizer response curves. *Proc. 9th Cong. Intl. Potash Instit.* (Antibes, 1970), pp. 461–73.

Boyd, D. A. (1973) Developments in field experimentation with fertilizers. *Phosphorus in Agric.* No. 61, 7–17.

Boyd, D. A. *et al.* (1970) Nitrogen requirements of sugar beet grown on mineral soils. *Jour. Agric. Sci., Camb.* **74**, 37–46.

Boyd, D. A., Tong Kwong Yuen, L. and Needham, P. (1976) Nitrogen requirement of cereals. Part 1. Attributes of different types of response functions: curves and split lines. Part 2. Multi-level nitrogen tests with cereals. *Jour. Agric. Sci., Camb.* (in press).

Bradley, R. A. and Hunter, J. S. (1958) Determination of optimum operating conditions by experimental methods. Part I: Mathematics and statistics fundamental to the fitting of response surfaces; Part II: Models and methods. *Industrial Quality Control* **14** (1), 16–20; (6), 16–24; (7), 7–15; (8), 6–14.

Brody, S. (1945) *Bioenergetics and Growth.* Reinhold, New York.

Brown, W. G. and Arscott, G. H. (1960) Animal production functions and optimum ration specifications. *Jour. Farm. Econ.* **42**, 69–78.

Brown, W. G. and Beattie, B. R. (1975) Improving estimates of economic parameters by use of ridge regression with production function applications. *Amer. Jour. Agric. Econ.* **57**, 21–32.

Brown, W. G. and Merril, M. O. (1958) Production functions from data over a series of years. *Jour. Farm Econ.* **40**, 451–7.

Brown, W. G., Jackson, T. L. and Petersen, R. G. (1962) A method for incorporating soil test measurement into fertilizer response functions. *Agronomy Jour.* **54**, 152–4.

Burt, O. R. (1963) Economic replacement. *S.I.A.M. Rev.* **5**, 203–8; **6**, 59.

Burt, O. R. and Stauber, M. S. (1971) Economic analysis of irrigation in subhumid climate. *Amer. Jour. Agric. Econ.* **53**, 33–46.

Byerlee, D. R. and Anderson, J. R. (1969) Value of predictors of uncontrolled factors in response functions. *Australian Jour. Agric. Econ.* **13**, 118–27.

Cady, F. B. and Laird, R. J. (1969) Bias error in yield function as influenced by treatment design and postulated model. *Soil Sci. Soc. Amer. Proc.* **33**, 282–6.

Campbell, N. A. and Keay, J. (1970) Flexible techniques in describing mathematically a range of response curves of pasture species. *Proc. 11th Intl. Grasslands Congress,* pp. 332–4.

Candler, W. V. (1962) Production economics and problems of animal production. *Proc. N.Z. Soc. Animal Prodn.* **22**, 142–58.

Candler, W. and Cartwright, W. (1970) Taxation: A neglected aspect of production economics. *Canadian Jour. Agric. Econ.* **18**, 76–89.

Cannon, D. J. (1969) The wool production and liveweight of wethers in relation to stocking rate and superphosphate application. *Australian Jour. Expt. Agric. Animal Husb.* **9**, 172–80.

Cannon, D. J. (1972) The influence of rate of stocking and application of superphosphate on the production and quality of wool, and on the gross margin from Merino wethers. *Australian Jour. Expt. Agric. Animal Husb.* **12**, 348–54.

Carley, D. H. (1973) Silage-concentrate substitution; Effects on milk production and income over feed cost. *Amer. Jour. Agric. Econ.* **55**, 641–6.

REFERENCES 181

CARRUTHERS, I. D. and DONALDSON, G. F. (1971) Estimation of effective risk reduction through irrigation of a perennial crop. *Jour. Agric. Econ.* **22**, 39–48.
CARTER, H. O., DEAN, G. W. and MCCORKLE, C. O. (1960) Economics of fertilization for selected California crops. Calif. Agric. Expt. Sta. Giannini Foundation, Mimeo Rept. 230, Berkeley.
CATE, R. B. (1976) Law of the minimum. *Encyclopedia of Earth Science*, Vol. VIB: *Soil Science*.
CHEW, V. (Ed.) (1958) *Experimental Designs in Industry*. Wiley, New York.
CHISHOLM, A. H. (1965) Towards the determination of optimum stocking rates in the high rainfall zone. *Rev. Mktng. Agric. Econ.* **33**, 5–31.
CHISHOLM, A. H. (1966) Criteria for determining the optimum replacement pattern. *Jour. Farm Econ.* **48**, 107–12.
CHISHOLM, A. H. (1974) Effects of tax depreciation policy and investment incentives on optimal equipment replacement decisions. *Amer. Jour. Agric. Econ.* **56**, 776–83.
CHISHOLM, A. H. (1975) Income taxes and investment decisions: The long-life appreciating asset case. *Economic Inquiry* **13**, 565–78.
CHISHOLM, A. H. and DILLON, J. L. (1966) *Discounting and Other Interest Rate Procedures in Farm Management*. Dept. Agric. Econ. and Bus. Mgmt., Univ. of New England (Farm Mgmt. Guidebook No. 2), Armidale.
CHOWDHURY, S. R., NAGADEVARA, V. and HEADY, E. O. (1975) A Bayesian application on Cobb-Douglas production function. *Amer. Jour. Agric. Econ.* **57**, 361–3.
CLYDE, H. S. et al. (1923) Economic use of irrigation water based on tests. *Engineering News Record* **91**, 548–52.
COCHRAN, W. G. (1973) Experiments for nonlinear functions. *Jour. Amer. Stat. Assocn.* **68**, 771–81.
COCHRAN, W. G. and COX, G. M. (1957) *Experimental Designs*, Revised edn. Wiley, New York.
COLWELL, J. D. (1967–8) Calibration and assessment of soil tests for estimating fertilizer requirements: I. Statistical models and tests of significance. II. Fertilizer requirements and an evaluation of soil testing. *Australian Jour. Soil Res.* **5**, 275–93; **6**, 93–103.
COLWELL, J. D. (1970) Precision in the estimation of the fertilizer requirements of wheat: Need, feasibility and economic desirability. *Jour. Australian Instit. Agric. Sci.* **36**, 273–8.
COLWELL, J. D. (1973a) The derivation of fertilizer recommendations for crops in a non-uniform environment. *Pontificae Academiae Scientiarum Scripta Varia* No. 38, 935–61.
COLWELL, J. D. (1973b) Assessments of the relative values of compound nitrogen-phosphorus fertilizers for wheat production. *Australian Jour. Agric. Econ.* **17**, 189–99.
COLWELL, J. D. (1976) *Computations for Studies of Soil Fertility*. Divn. of Soils, C.S.I.R.O., Canberra.
COLWELL, J. D. and ESDAILE, R. J. (1966) The application of production function analysis for the estimation of fertilizer requirements of wheat in northern New South Wales. *Australian Jour. Expt. Agric. Animal Husb.* **6**, 418–24.
COLWELL, J. D. and ESDAILE, R. J. (1968) The calibration, interpretation and evaluation of tests for the phosphorus fertilizer requirements of wheat in northern N.S.W. *Australian Jour. Soil Res.* **6**, 105–20.
COLWELL, J. D. and STACKHOUSE, K. M. (1970) Some problems in the estimation of simultaneous fertilizer requirements of crops from response surfaces. *Australian Jour. Expt. Agric. Animal Husb.* **10**, 183–95.

COLYER, D. (1969) Fertilization strategy under uncertainty. *Canadian Jour. Agric. Econ.* **17**, 144–9.
COLYER, D. and KROTH, E. M. (1968) Corn yield response and economic optima for nitrogen treatments and plant population over a seven-year period. *Agronomy Jour.* **60**, 524–9.
COLYER, D. and KROTH, E. M. (1970) Expected yields and returns for corn due to nitrogen and plant population. *Agronomy Jour.* **62**, 487–90.
CONE, B. W. (1974–5) The riskiness of adopting the use of fertilizer – a Brazilian example. *Intl. Jour. Agrarian Affairs*, Suppl., pp. 96–109.
CONNIFFE, D. *et al.* (1970) Experimental design for grazing trials. *Jour. Agric. Sci., Camb.* **74**, 339–42.
CONNIFFE, D. *et al.* (1972) An example of a method of statistical analysis of a grazing trial. *Jour. Agric. Sci., Camb.* **79**, 165–7.
COWLISHAW, S. J. (1969) The carrying capacity of pastures. *Jour. Brit. Grassland Soc.* **24**, 207–14.
DE DATTA, S. K. and BARKER, R. (1968) Economic analysis of experimental results in rice production. *Philippine Econ. Jour.* **7**, 162–83.
DE JANVRY, A. (1972a) Optimal levels of fertilization under risk: The potential for corn and wheat fertilization under alternative price policies in Argentina. *Amer. Jour. Agric. Econ.* **54**, 1–10.
DE JANVRY, A. (1972b) The generalized power production function. *Amer. Jour. Agric. Econ.* **54**, 234–7.
DE OLIVEIRA, A. J. (1973) Análise econometrica da experimentacão de fertilizantes no trigo cultivado no Alentejo (Portugal). *Agronomia Lusitana* **34**, 5–175.
DANO, S. (1966) *Industrial Production Models.* Springer-Verlag, Vienna.
DAVIDSON, B. R. (1962) Crop yields in experiments and on farms. *Nature* **194**, 458–9.
DAVIDSON, B. R. and MARTIN, B. R. (1965a) The relationship between yields on farms and in experiments. *Australian Jour. Agric. Econ.* **9**, 129–40.
DAVIDSON, B. R. and MARTIN, B. R. (1965b) The use of experimental results in farm planning. *Farm Policy* (Univ. of Western Australia) **4**, 109–14.
DAVIDSON, B. R. and MARTIN, B. R. (1967) *The Relationship between Farm and Experimental Yields.* Univ. of Western Australia Press, Nedlands.
DAVIDSON, B. R., MARTIN, B. R. and MAULDON, R. G. (1967) The application of experimental research to farm production. *Jour. Farm Econ.* **49**, 900–7.
DAVIS, J. F., SUNDQUIST, W. B. and FRAKES, M. G. (1959) The effect of fertilizers on sugar beets including an economic optima study of the response. *Jour. Amer. Soc. Sugar Beet Technologists* **10**, 424–34.
DAY, R. H. (1965) Probability distributions of field crop yields. *Jour. Farm. Econ.* **47**, 713–41.
DEAN, G. W. (1960) Consideration of time and carryover effects in milk production functions. *Jour. Farm Econ.* **42**, 1512–4.
DEAN, G. W. *et al.* (1972) Production functions and linear programming models for dairy cattle feeding. *Calif. Agric. Expt. Sta. Giannini Foundation Monograph* **31**, Berkeley.
DENMEAD, O. T. and SHAW, R. H. (1962) Availability of soil water to plants as affected by soil moisture content and meteorological conditions. *Agronomy Jour.* **54**, 385–90.
DENT, J. B. (1964) Optimal rations for livestock with special reference to bacon pigs. *Jour. Agric. Econ.* **16**, 68–87.

DENT, J. B. (1966) Towards the ideal in pig ration formulation. *Proc. 2nd Pig Industry Developmnt. Assocn. Conf.* (Brighton), pp. 58–71.
DENT, J. B. et al. (1970) Protein lysine and feed intake level effects on pig growth. III. Regression analysis and economic effects. *Journ. Agric. Sci., Camb.* **75**, 189–205.
DENT, J. B. and ANDERSON, J. R. (Eds.) (1971) *Systems Analysis in Agricultural Management.* Wiley, Sydney.
DENT, J. B. and ENGLISH, P. R. (1966) The evaluation of economically optimal rations for bacon pigs formulated by curve fitting and linear programming techniques. *Animal Prodn.* **8**, 213–20.
DENT, J. B., ENGLISH, P. R. and RAEBURN, J. R. (1970) Testing regression models of pig feeding systems. *Animal Prodn.* **12**, 379–92.
DHRYMES, P. J. (1970) *Econometrics: Statistical Foundations and Applications.* Harper and Row, New York.
DIAMOND, P. A. and STIGLITZ, J. E. (1974) Increases in risk and in risk aversion. *Jour. Econ. Theory* **8**, 337–60.
DILLON, J. L. (1966) Economic considerations in the design and analysis of agricultural experiments. *Rev. Mktng. Agric. Econ.* **34**, 64–75.
DILLON, J. L. (1971) An expository review of Bernoullian decision theory: Is utility futility? *Rev. Mktng. Agric. Econ.* **39**, 3–80.
DILLON, J. L. (1975) The economics of systems research. *Agric. Systems* **1**, 5–22.
DILLON, J. L. and ANDERSON, J. R. (1971) Allocative efficiency, traditional agriculture and risk. *Amer. Jour. Agric. Econ.* **53**, 26–32.
DILLON, J. L. and BURLEY, H. T. (1961) A note on the economics of grazing and its experimental investigation. *Australian Jour. Agric. Econ.* **5**, 123–32.
DILLON, J. L. and OFFICER, R. R. (1971) Economic v. statistical significance in agricultural research and extension. *Farm Economist* **12**, 1–15.
DOLL, J. P. (1959) The allocation of limited quantities of variable resources among competing farm enterprises. *Jour. Farm Econ.* **41**, 781–9.
DOLL, J. P. (1967) An analytical technique for estimating weather indexes from meteorological measurements. *Jour. Farm Econ.* **49**, 79–88.
DOLL, J. P. (1971) Obtaining preliminary Bayesian estimates of the value of a weather forecast. *Amer. Jour. Agric. Econ.* **53**, 651–5.
DOLL, J. P. (1972) A comparison of annual versus average optima for fertilizer experiments. *Amer. Jour. Agric. Econ.* **54**, 226–33.
DOLL, J. P., JEBE, E. H. and MUNSON, R. D. (1960) Computation of variance estimates for marginal physical products and marginal rates of substitution. *Jour. Farm Econ.* **42**, 596–607.
DORFMAN, R. (1963) Response of agricultural yields to water in the former Punjab. In White House Panel on Water Logging and Salinity in West Pakistan, *Report on Land and Water Development in the Indus Plain*, The White House, Washington, pp. 417–37.
DORFMAN, R. (1964) *The Price System.* Prentice-Hall, Englewood Cliffs.
DOWDLE, B. (1962) Investment theory and forest management planning. *Yale Univ. School of Forestry Bul.* **67**, New Haven.
DOWNEY, L. A. (1972) Water-yield relations for non-forage crops. *Jour. Irrign. Drainage Divn., Amer. Soc. Civil Eng.* **98**, 107–15.
DRAPER, N. R. (1961) Missing values in response surface designs. *Technometrics* **3**, 389–98.
DRAPER, N. R. (1963) Ridge analysis of response surfaces. *Technometrics* **5**, 469–80.
DRAPER, N. R. and SMITH, H. (1966) *Applied Regression Analysis.* Wiley, New York.

DUDLEY, N. J. and BURT, O. R. (1973) Stochastic reservoir management and system design for irrigation. *Water Resources Res.* **9**, 507–22.

DUDLEY, N. J., HOWELL, D. T. and MUSGRAVE, W. F. (1971) Irrigation planning. 2: Choosing optimal acreages within a season. *Water Resources Res.* **7**, 1051–63.

DULOY, J. (1959) Resource allocation and the fitted production function. *Australian Jour. Agric. Econ.* **3** (2), 75–85.

DULOY, J. and BATTESE, G. E. (1967) Time and recursiveness in livestock feeding trials. *Australian Jour. Agric. Econ.* **11**, 184–91.

DYCKMAN, T. R., SMIDT, S. and MCADAMS, A. K. (1969) *Management Decision Making under Uncertainty.* Collier–Macmillan, London.

EIDMAN, V. R., LINGLE, J. C. and CARTER, H. O. (1963) Optimum fertilization rates for crops with multi-harvest periods. *Jour. Farm Econ.* **45**, 823–30.

ELANDT, R. C. (1963) Optimal and sufficient allocation of multiple varietal experiments. *Biometrics* **19**, 615–28.

ENGELSTAD, O. P. (1963) Effect of variation in fertilizer rates and ratios on yield and profit surfaces. *Agronomy Jour.* **55**, 263–5.

ENGELSTAD, O. P. and DOLL, E. C. (1961) Corn yield response to applied phosphorus as affected by rainfall and temperature variables. *Agronomy Jour.* **53**, 389–92.

ENGELSTAD, O. P. and PARKS, W. L. (1971) Variability in optimum N rates for corn. *Agronomy Jour.* **63**, 21–3.

EVENSON, R. E. and KISLEV, Y. (1975) *Agricultural Research and Productivity.* Yale Univ. Press, New Haven.

EWALT, R. L., DOLL, J. P. and DECKER, W. L. (1961) Correlation of drouth indices with corn yields. *Univ. of Missouri Agric. Expt. Stat. Res. Bul.* 788, Columbia.

EZEKIEL, M. and FOX, K. A. (1959) *Methods of Correlation and Regression Analysis.* Wiley, New York.

FARIS, J. E. (1960a) Analytical techniques used in determining the optimum replacement pattern. *Jour. Farm Econ.* **42**, 755–66.

FARIS, J. E. (1960b) Economics of replacing cling peach trees. Calif. Agric. Expt. Sta. Giannini Foundation, Mimeo Rept. 232, Berkeley.

FARIS, J. E. (1961) On determining the optimum replacement pattern: A reply. *Jour. Farm Econ.* **43**, 952–5.

FAWCETT, R. H. (1973) Towards a dynamic production function. *Jour. Agric. Econ.* **24**, 543–55.

FEDERER, W. T. and RAGHAVARAO, D. (1975) On augmented designs. *Biometrics* **31**, 29–35.

FINNEY, D. J. (1960) *An Introduction to the Theory of Experimental Design.* Chicago Univ. Press.

FISHER, R. A. (1924) The influence of rainfall on the yield of wheat at Rothamsted. *Phil. Trans. Roy. Soc. London* **B.213**, 89–142.

FITTS, J. W. et al. (1959) Determining yield response surfaces and economically optimum fertilizer rates for corn under various soil and climatic conditions in North Carolina. TVA Rept. Contract TV-13434A Project N.C. 863, Knoxville.

FLINN, J. C. and MUSGRAVE, W. F. (1967) Development and analysis of input–output relations for irrigation water. *Australian Jour. Agric. Econ.* **11**, 1–19.

FOX, K. A. (1968) *Intermediate Economic Statistics.* Wiley, New York.

FRIED, M. and BROESHART, H. (1967) *The Soil–Plant System.* Academic Press, New York.

FRISCH, R. (1965) *Theory of Production.* Reidel, Dordrecht.

FULLER, W. A. (1962) Estimating the reliability of quantities derived from empirical production functions. *Jour. Farm Econ.* **44**, 82–99.
FULLER, W. A. (1965) Stochastic fertilizer production functions for continuous corn. *Jour. Farm Econ.* **47**, 105–19.
FULLER, W. A. (1969) Grafted polynomials as approximating functions. *Australian Jour. Agric. Econ.* **13**, 35–46.
FULLER, W. A. (1976) *Introduction to Statistical Time Series.* Wiley, New York.
FULLER, W. A. and BATTESE, G. E. (1973) Transformations for estimation of linear models with nested-error structure. *Jour. Amer. Stat. Assocn.* **68**, 626–32.
FULLER, W. A. and BATTESE, G. E. (1974) Estimation of linear models with crossed-error structure. *Jour. Econometrics* **2**, 67–78.
FULLER, W. A. and CADY, F. B. (1965) Estimation of asymptotic rotation and nitrogen effects. *Agronomy Jour.* **57**, 299–302.
GAFFNEY, M. M. (1960) Concepts of financial maturity in timber. *North Carolina State College A.E. Informn. Ser.* No. 62, Raleigh.
GAFFNEY, M. M. (1967) Tax-induced slow turnover of capital. *Western Econ. Jour.* **5**, 308–23.
GAFFNEY, M. M. (1970–1) Tax-induced slow turnover of capital. *Amer. Jour. Econ. Sociol.* **29**, 25–32, 179–97, 277–87, 409–24; **30**, 105–11.
GALLANT, A. R. and FULLER, W. A. (1973) Fitting segmented polynomial regression models whose join points have to be estimated. *Jour. Amer. Stat. Assocn.* **68**, 144–8.
GANE, M. (Ed.) (1968) *Martin Faustmann and the Evolution of Discounted Cash Flow.* Cwlth. Forestry Instit., Oxford.
GASTAL, E. (Ed.) (1971) *Analisis Economico de los Datos de la Investigacion en Ganaderia.* Instituto Interamericano de Ciencias Agricolas-Zona Sur-OEA, Montevideo.
GNANADESIKAN, R. (1963) Multivariate statistical methods for analysis of experimental data. *Industrial Quality Control* **19** (9), 22–32.
GODDEN, D. P. and HELYAR, K. R. (1975) Optimality, hurrah! *Jour. Australian Instit. Agric. Soc.* **41**, 197–8.
GODDEN, D. P. and HELYAR, K. R. (1976) A modified theory for calculating optimal fertilizer rates. *Australian Jour. Agric. Econ.* **20**, (in press).
GREIG, I. D. (1972) Beef production models in management systems. *Proc. Australian Soc. Animal Prodn.* **9**, 89–93.
HADAR, J. (1965) Optimal resource allocation in broiler production. *Farm Economist* **10**, 359–66.
HADAR, J. and RUSSELL, W. R. (1969) Rules for ordering uncertain prospects. *Amer. Econ. Rev.* **59**, 25–34.
HADAR, J. and RUSSELL, W. R. (1974) Stochastic dominance in choice under uncertainty. In M. Balch and D. McFadden (Eds.), *Essays on Economic Behavior under Uncertainty*, North-Holland, Amsterdam, pp. 133–50.
HADLEY, G. (1967) *Introduction to Probability and Statistical Decision Theory.* Holden-Day, San Francisco.
HALL, W. A. and BURAS, N. (1961) Optimum irrigation practice under conditions of deficient water supply. *Transactns. Amer. Soc. Agric. Engineers* **4**, 131–4.
HALL, W. A., ASCE, M. and BUTCHER, W. S. (1968) Optimal timing of irrigation. *Jour. Irrign. Drainage Divn., Amer. Soc. Civil Eng.* **94**, 267–75.
HALL, W. B. (1975) Repeated measurements experiments. In V. J. Bofinger and J. L. Wheeler (Eds.), *Developments in Field Experiment Design and Analysis*, Cwlth. Agric. Bur., Slough, pp. 33–41.

HALTER, A. N. (1963) A review of decision-making literature with a view of possibilities for research on decision-making processes of western ranchers. In Western Agric. Econ. Res. Council Cmtee. on Econ. of Range Use, *Development and Evaluation of Research in Range Management Decision Making*, Rept. No. 5 [Berkeley].

HALTER, A. N. and DEAN, G. W. (1971) *Decisions under Uncertainty with Research Applications*. South Western, Cincinnati.

HALTER, A. N., CARTER, H. O. and HOCKING, J. G. (1957) A note on the transcendental production function. *Jour. Farm. Econ.* **39**, 966–74.

HAMILTON, D. and BATH, J. G. (1970) Performance of sheep and cattle grazed separately and together. *Australian Jour. Expt. Agric. Animal Husb.* **10**, 19–26.

HANSEN, P. L. and MIGHELL, R. L. (1956) Economic choices in broiler production. *U.S.D.A. Tech. Bul.* 1154, Washington.

HARLAN, J. R. (1958) Generalized curves for gain per head and gain per acre in rates of grazing studies. *Jour. Range Mgmt.* **11**, 140–7.

HART, R. H. (1972) Forage yield, stocking rate, and beef gains on pasture. *Herbage Abstracts* **42**, 345–53.

HARTLEY, H. O. (1964) Experimental designs for estimating the characteristics of response functions. In O.E.C.D., *Co-operation between Research in Agricultural Natural Sciences and Agricultural Economics* (O.E.C.D. Documentation in Food and Agriculture No. 65), Paris, pp. 163–76.

HAVLICEK, J. and SEAGRAVES, J. A. (1962) The cost of the wrong decision as a guide in production research. *Jour. Farm Econ.* **44**, 157–68.

HAZELL, P. B. R. and SCANDIZZO, P. L. (1975) Market intervention policies when production is risky. *Amer. Jour. Agric. Econ.* **57**, 641–9.

HEADY, E. O. (1952) *Economics of Agricultural Production and Resource Use*. Prentice-Hall, Englewood Cliffs.

HEADY, E. O. (1964) Review of the present situation in North America. In O.E.C.D., *Co-operation between Research in Agricultural Natural Sciences and Agricultural Economics* (O.E.C.D. Documentation in Food and Agriculture No. 65), Paris, pp. 45–107.

HEADY, E. O. (1966) *Agricultural Problems and Policies of Developed Countries*. Bondenes Forlag, Oslo.

HEADY, E. O. et al. (1963a) Beef cattle production functions in forage utilization. *Iowa State Univ. Agric. Expt. Sta. Res. Bul.* 517, Ames.

HEADY, E. O. et al. (1963b) Experimental production and profit functions for beef steers. *Canadian Jour. Agric. Econ.* **11**, 29–40.

HEADY, E. O. et al. (1964a) Milk production functions incorporating variables for cow characteristics and environment. *Jour. Farm. Econ.* **46**, 1–19.

HEADY, E. O. et al. (1964b) Milk production functions in relation to feed inputs, cow characteristics and environmental conditions. *Iowa State Univ. Agric. Expt. Sta. Res. Bul.* 529, Ames.

HEADY, E. O. and CANDLER, W. V. (1958) *Linear Programming Methods*. Iowa State Univ. Press, Ames.

HEADY, E. O. and DILLON, J. L. (1961) *Agricultural Production Functions*. Iowa State Univ. Press, Ames.

HEADY, E. O. and PESEK, J. (1955) Hutton and Thorne on isoclines: A reply. *Jour. Farm Econ.* **37**, 363–8.

HEADY, E. O. and TWEETEN, L. G. (1963) *Resource Demand and Structure of the Agricultural Industry*. Iowa State Univ. Press, Ames.

HEADY, E. O., PESEK, J. and MCCARTHY, W. O. (1963) Production functions and methods of specifying optimum fertilizer use under various uncertainty conditions for hay. *Iowa State Univ. Agric. Expt. Sta. Res. Bul.* 518, Ames.

HEADY, E. O., PESEK, J. and RAO, V. Y. (1966) Fertilizer production functions from experimental data with associated supply and demand relationships. *Iowa State Univ. Agric. Expt. Sta. Res. Bul.* 543, Ames.

HELYAR, K. R. and GODDEN, D. P. (1976) The biology and modelling of fertilizer response. *Jour. Australian Instit. Agric. Sci.* **41**, (in press).

HENDERSON, J. M. and QUANDT, R. E. (1971) *Microeconomic Theory: A Mathematical Approach.* McGraw-Hill, New York.

HERDT, R. W. and BERNSTEN, R. H. (1975) Methodology for assessing yield constraints. *Intl. Rice Res. Instit. IRAEN Working Paper 2*, Manila.

HERDT, R. W., DE DATTA, S. K. and NEELEY, D. (1975) Farm yield constraints in Nueva Ecija and Laguna, Philippines, 1974. *Intl. Rice Res. Instit. IRAEN Working Paper 3*, Manila.

HIEBERT, L. D. (1974) Risk, learning and the adoption of fertilizer responsive seed varieties. *Amer. Jour. Agric. Econ.* **56**, 764–8.

HILDRETH, C. G. (1957) Possible models for agronomic-economic research. In E. L. Baum et al. (Eds.), *Economic and Technical Analysis of Fertilizer Innovations and Resource Use*, Iowa State Univ. Press, Ames, pp. 176–86.

HILDRETH, R. J. (1957) Influence of rainfall on fertilizer profits. *Jour. Farm Econ.* **39**, 522–4.

HILL, W. J. and HUNTER, W. G. (1966) A review of response surface methodology: A literature survey. *Technometrics* **8**, 571–90.

HILL, W. J. and HUNTER, W. G. (1974) Design of experiments for subsets of parameters. *Technometrics* **16**, 425–34.

HIRSHLEIFER, J. (1971) The private and social value of information and the reward to inventive activity. *Amer. Econ. Rev.* **61**, 561–74.

HJELM, L. J. (1962) Demonstration of input–output data obtained from technical experiments with special reference to fertilizers and livestock nutrition. In O.E.C.D., *Inter-disciplinary Co-operation in Technical and Economic Agricultural Research* (O.E.C.D. Documentation in Food and Agriculture No. 50), Paris, pp. 115–21.

HOCHMAN, E. and LEE, I. M. (1972) Optimal decision in the broiler producing firm: A problem of growing inventory. *Calif. Agric. Expt. Sta. Giannini Foundation Monograph* 29, Berkeley.

HOEL, P. G. (1964) Methods for comparing growth type curves. *Biometrics* **20**, 859–72.

HOEPNER, P. H. and FREUND, R. J. (1964) A methodological approach to the estimation of time-quantity broiler production functions. *Virginia Polytechnic Instit. Agric. Expt. Sta. Tech. Bul.* 170, Blacksburg.

HOERL, A. E. and KENNARD, R. W. (1970a) Ridge regression: Biased estimation for nonorthogonal problems. *Technometrics* **12**, 55–67.

HOERL, A. E. and KENNARD, R. W. (1970b) Ridge regression: Applications to non-orthogonal problems. *Technometrics* **12**, 69–82.

HOFFNAR, B. R. (1963) Some practical statistical notions for production function studies. *Jour. Farm Econ.* **45**, 1226–31.

HOFFNAR, B. R. and JOHNSON, G. L. (1966) Summary and evaluation of the cooperative agronomic-economic experimentation at Michigan State University. *Michigan State Univ. Expt. Sta. Res. Bul.* 11, East Lansing.

Hogg, H. C. et al. (1969) Economics of a water-yield function for sugar cane. *Jour. Irrign. Drainage Divn.*, *Amer. Soc. Civil Eng.* **95**, 127–38.

Hoglund, C. R. et al. (Eds.) (1959) *Nutritional and Economic Aspects of Feed Utilization by Dairy Cows*. Iowa State Univ. Press, Ames.

Holder, J. M., Wilson, B. R. and Williams, R. J. (1969) A response surface approach to examining the use of separated milk and wheat by growing pigs. *Australian Jour. Expt. Agric. Animal Husb.* **9**, 121–6.

Hoover, L. M. et al. (1967) Economic relationships of hay and concentrate consumption to milk production. *Jour. Farm. Econ.* **49**, 64–78.

Huang, D. S. (1970) *Regression and Econometric Methods*. Wiley, New York.

Hunter, W. G., Hill, W. J. and Henson, T. L. (1969) Designing experiments for precise estimation of all or some of the constants in a mechanistic model. *Canadian Jour. Chem. Eng.* **47**, 76–85.

Hutton, R. F. (1955) Further comment on the Heady-Pesek fertilizer production function. *Jour. Farm Econ.* **37**, 566–8.

Hutton, R. F. and Thorne, D. W. (1955) Review notes on the Heady-Pesek fertilizer production surface. *Jour. Farm Econ.* **37**, 117–9.

I.R.R.I. (1969/70/72) *Annual Report*. Intl. Rice Res. Instit., Los Baños.

Inkson, R. H. E. (1964) The precision of estimates of the soil content of phosphate using the Mitscherlich response equation. *Biometrics* **20**, 873–82.

Jardine, R. (1975a) Two cheers for optimality! *Jour. Australian Instit. Agric. Sci.* **41**, 30–4.

Jardine, R. (1975b) Rejoinder. *Jour. Australian Instit. Agric. Sci.* **41**, 199–201.

Jensen, D. and Pesek, J. (1959) Generalization of yield equations in two or more variables. *Agronomy Jour.* **51**, 255–63.

Johnson, N. L. and Leone, F. C. (1964) *Statistics and Experimental Design in Engineering and the Physical Sciences*, 2 vols., Wiley, New York.

Johnson, P. R. (1953) Alternative functions for analyzing a fertilizer-yield relationship. *Jour. Farm Econ.* **35**, 519–29.

Johnson, R. W. M. (1971) Aggregation of micro-functions to obtain a whole farm production function. *Australian Jour. Agric. Econ.* **15**, 151–60.

Jones, R. B. and Hocknell, K. E. (1962) The economics of fertilizer application to permanent pastures for beef production. *Univ. of Nottingham Dept. of Agric. Econ. F.R.* 147, Sutton Bonnington.

Jones, R. J. and Sandland, R. L. (1974) The relation between animal gain and stocking rate: Derivation of the relation from the results of grazing trials. *Jour. Agric. Sci., Camb.* **83**, 335–42.

Jonsson, L. (1974) On the choice of a production function model for nitrogen fertilization on small grain farms in Sweden. *Swedish Jour. Agric. Res.* **4**, 87–97.

Kaminsky, M. (1974) Estimacion de hipersuperficies de produccion de producto multiple. *Cuadernos de Economia* **11**, 65–89.

Kennard, R. W. (1963) Statistics and technology. *Industrial Quality Control* **19** (12), 21–3.

Kennedy, J. O. S. (1972) A model for determining optimal marketing and feeding policies for beef cattle. *Jour. Agric. Econ.* **23**, 147–59.

Kennedy, J. O. S. (1973) Control systems in farm planning. *European Rev. Agric. Econ.* **1**, 415–33.

Kennedy, J. O. S. et al. (1973) Optimal fertilizer carryover and crop recycling policies or a tropical grain crop. *Australian Jour. Agric. Econ.* **17**, 104–13.

KENNEDY, J. O. S. et al. (1976) Optimal feeding policies for broiler production: An application of dynamic programming. *Australian Jour. Agric. Econ.* **20**, 19–32.
KENWORTHY, O. O. (1963) Factorial experiments with mixtures using ratios. *Industrial Quality Control* **19** (12), 24–6.
KMENTA, J. (1971) *Elements of Econometrics.* Collier-Macmillan, London.
KNETSCH, J. L. (1959) Moisture uncertainties and fertility response studies. *Jour. Farm Econ.* **41**, 70–6.
KUPPER, L. L. and MEYDRECH, E. F. (1974) Experimental design considerations based on a new approach to mean square error estimation of response surfaces. *Jour. Amer. Stat. Assocn.* **69**, 461–3.
LAIRD, R. J. et al. (1969) Combining data from fertilizer experiments into a function useful for estimating specific fertilizer recommendations. *CIMMYT Res. Bul.* 12, Mexico City.
LAIRD, R. J. and CADY, F. B. (1969) Combined analysis of yield data from fertilizer experiments. *Agronomy Jour.* **61**, 829–34.
LEFTWICH, R. H. (1970) *The Price System and Resource Allocation*, 4th edn. Dryden Press, Hinsdale.
LIEBIG, J. von (1855) *Die Grundsätze der Agriculturchemie.* Viewig und Sohn, Braunschweig.
LIN, C. Y. (1974) Numerical techniques for evaluating sample information. *Technometrics* **16**, 447–54.
LINDLEY, D. V. (1965) *Introduction to Probability and Statistics from a Bayesian Viewpoint*, 2 vols. Cambridge Univ. Press.
LLOYD, A. G. (1966) Economic aspects of stocking and feeding policies in the sheep industry in southern Australia. *Proc. Australian Soc. Animal Prodn.* **6**, 137–47.
LOVELL, A. C. et al. (1974) Estimating the response of feedlot beef cattle to various corn silage rations. *Canadian Jour. Agric. Econ.* **22**, 53–7.
LUCAS, J. M. (1974) Optimum composite designs. *Technometrics* **16**, 561–7.
MCARTHUR, I. D. (1970) Optimum sheep stocking rate. *Jour. Australian Institute Agric. Sci.* **36**, 9–14.
MCARTHUR, I. D. and DILLON, J. L. (1971) Risk, utility and stocking rate. *Australian Jour. Agric. Econ.* **15**, 20–35.
MCCALL, J. J. (1971) Probabilistic microeconomics. *Bell Jour. Econ. Mgmt. Sci.* **2**, 403–33.
MCCONNEN, R. J. (1965) Relation between the pattern of use and the future output from a flow resource. *Jour. Farm Econ.* **47**, 311–23.
MCCONNEN, R. J. et al. (1963) Feed-livestock relationships — a model for analysing management decisions. *Agric. Econ. Res.* **15**, 41–8.
MAGNUSSON, G. (1969) *Production under Risk: A Theoretical Study.* Almqvist and Wiksells, Uppsala.
MANDAC, A. M. (1974) An economic analysis of factors affecting yield of rice in 1973–74 Central Luzon farmers' field experiments. *Intl. Rice Res. Instit. Dept. Agric. Econ. Paper* No. 74–24, Los Baños.
MARQUARDT, D. W. (1966) *Least Squares Estimation of Nonlinear Parameters.* IBM Share Library (Dist. No. 309401), New York.
MARQUARDT, D. W. (1970) Generalized inverses, ridge regression, biased linear estimation and non-linear estimation. *Technometrics* **12**, 591–612.
MARQUARDT, D. W. (1975) Ridge regression in practice. *Amer. Statistician* **29**, 3–20.
MASON, D. D. (1956) Functional models and experimental designs for characterizing response curves and surfaces. In E. L. Baum et al. (Eds.), *Economic Analysis of Fertilizer Use Data*, Iowa State Univ. Press, Ames, pp. 76–98.

MAULDON, R. G. (1968) Economic implications of stocking rate experiments. *Proc. Australian Grasslands Conf.* Vol. 2, pp. 158–88.

MENDENHALL, W. (1968) *Introduction to Linear Models and the Design and Analysis of Experiments.* Wadsworth, Belmont.

MIHRAM, G. A. (1972) *Simulation: Statistical Foundations and Methodology.* Academic Press, New York.

MILHORN, H. T. (1966) *Application of Control Theory to Physiological Systems.* Saunders, Philadelphia.

MINHAS, B. S., PARIKH, K. S. and SRINIVASAN, T. N. (1974) Toward the structure of a production function for wheat yields with dated inputs of irrigation water. *Water Resources Res.* **10**, 383–93.

MONTAÑO, C. B. and BARKER, R. (1970) The effect of solar energy on rice yield response to nitrogen. *Intl. Rice Res. Instit. Dept. Agric. Econ. Paper* No. C.2(70), Los Baños.

MONTERO, E. and PEREZ, S. (Eds.) (1967) *Investigacion Economica y Experimentacion Agricola.* Instituto Interamericano de Ciencias Agricolas-Zona Sur-OEA, Montevideo.

MOORE, C. V. (1961) A general analytical framework for estimating the production function for crops using irrigation water. *Jour. Farm Econ.* **43**, 876–88.

MORLEY, F. H. W. and SPEDDING, C. R. W. (1968) Agricultural systems and grazing experiments. *Herbage Abstracts* **38**, 279–87.

MUNDLAK, Y. (1964) Transcendental multiproduct production functions. *Intl. Econ. Rev.* **5**, 273–84.

MUNSON, R. D. and DOLL, J. P. (1959) Economics of fertilizer use in crop production. *Advances in Agronomy* **11**, 133–69.

MYERS, R. H. (1971) *Response Surface Methodology.* Allyn and Bacon, Boston.

MYERS, R. H. and CARTER, W. H. (1973) Response surface techniques for dual response systems. *Technometrics* **15**, 301–17.

NATL. RES. COUNCIL (1961) Status and methods of research in economic and agronomic aspects of fertilizer response and use. *Natl. Acad. Sci., Natl. Res. Council Publ.* **918**, Washington.

NELDER, J. A. (1966) Inverse polynomials, a useful group of multifactor response functions. *Biometrics* **22**, 128–41.

NELSON, M. and CASTLE, E. N. (1958) Profitable use of fertilizer on native meadows. *Jour. Range Mgmt.* **11**, 80–3.

NELSON, M., CASTLE, E. N. and BROWN, W. G. (1957) Use of the production function and linear programming in valuation of intermediate products. *Land. Econ.* **33**, 257–61.

NEWMAN, P. (1965) *The Theory of Exchange.* Prentice-Hall, Englewood Cliffs.

NIX, H. A. and FITZPATRICK, E. A. (1969) An index of crop water stress related to wheat and grain sorghum yields. *Agric. Meteorology* **6**, 321–37.

O.E.C.D. (1962) *Inter-disciplinary Co-operation in Technical and Economic Agricultural Research* (O.E.C.D. Documentation in Food and Agriculture No. 50), Paris.

O.E.C.D. (1964) *Co-operation between Research in Agricultural Natural Sciences and Agricultural Economics* (O.E.C.D. Documentation in Food and Agriculture No. 65), Paris.

O.E.C.D. (1965) *Co-operative Research to Improve Input/Output Data in Cow Milk Production* (O.E.C.D. Documentation in Food and Agriculture No. 71), Paris.

O.E.C.D. (1968a) *Co-operative Research in Input/Output Relationships in Poultry* (O.E.C.D. Documentation in Food and Agriculture No. 81), Paris.

O.E.C.D. (1968b) *Co-operative Research in Input/Output Relationships in Beef Production* (O.E.C.D. Documentation in Food and Agriculture No. 82), Paris.
O.E.C.D. (1969a) *Co-operative Research in Input/Output Relationships in Cow Milk Production* (O.E.C.D. Documentation in Food and Agriculture No. 83), Paris.
O.E.C.D. (1969b) *Co-operative Research in Input/Output Relationships in the Use of Fertilizers in Crop Production* (O.E.C.D. Documentation in Food and Agriculture No. 84), Paris.
OGUNFOWORA, O. and NORMAN, D. W. (1973) Farm-firm normative fertilizer demand response in the North Central State of Nigeria. *Jour. Agric. Econ.* **24**, 301-9.
OSTLE, B. (1963) *Statistics in Research*. Iowa State Univ. Press, Ames.
OURY, B. (1965) Allowing for weather in crop production model building. *Jour. Farm Econ.* **47**, 270-83.
OWEN, J. B. and RIDGMAN, W. J. (1968) The design and interpretation of experiments to study animal production from grazed pasture. *Jour. Agric. Sci., Camb.* **71**, 327-35.
PALTRIDGE, G. W. (1970) A model of the growing pasture. *Agric. Meteorology* **7**, 93-130.
PARIS, Q. *et al.* (1970) A note on milk production functions. *Amer. Jour. Agric. Econ.* **52**, 594-8.
PARKS, W. L. and KNETSCH, J. L. (1959) Corn yields as influenced by nitrogen level and drouth intensity. *Agronomy Jour.* **51**, 363-4.
PARKS, W. L. and KNETSCH, J. L. (1960) Utilizing drought-days in evaluating irrigation and fertility response studies. *Soil Sci. Soc. Amer. Proc.* **24**, 289-93.
PEQUIGNOT, R. and RECAMIER, A. (1962) Recherche d'un seuil de rentabilité pour la fumure azotée. *Bul. Technique d'Information des Ingenieurs des Services Agricoles* No. 175, 1-12.
PERRIN, R. K. (1972) Asset replacement principles. *Amer. Jour. Agric. Econ.* **54**, 60-7.
PERRIN, R. K. (1976) The value of information and the value of theoretical models in crop response research. *Amer. Jour. Agric. Econ.* **58**, 54-61.
PESEK, J. (1973) Crop yield response equations and economic levels of fertilizer use. *Pontificae Academiae Scientiarum Scripta Varia* No. 38, 881-925.
PETERSEN, R. G., LUCAS, H. L. and MOTT, G. O. (1965) Relationship between rate of stocking and per animal and per acre performance on pasture. *Agronomy Jour.* **57**, 27-30.
PRATT, J., RAIFFA, H. and SCHLAIFER, R. (1965) *Introduction to Statistical Decision Theory*. McGraw-Hill, New York.
PRESCOTT, J. A. (1928) Law of diminishing returns in agricultural experiment. *Econ. Record* **4**, 85-9.
RAIFFA, H. and SCHLAIFER, R. (1961) *Applied Statistical Decision Theory*. Harvard Business School, Boston.
RAM, M. (1963) Revision of economic water duty. In Intl. Comm. on Irrig. and Drainage, *Fifth Congress on Irrigation and Drainage*, New Delhi, pp. 15.513-32.
RAO, P. and MILLER, R. L. (1971) *Applied Econometrics*. Wadsworth, Belmont.
RAO, V. Y. and SENGUPTA, J. K. (1965) An econometric application of a generalized Cobb-Douglas production function. *Arthaniti* **8** (2), 1-12.
REID, G. K. R. and THOMAS, D. A. (1973) Pastoral production, stocking rate and seasonal conditions. *Qrtly. Rev. Agric. Econ.* **16**, 217-27.
RICHARDSON, A. E. V. and FRICKE, E. F. (1931) Effect of nitrogenous fertilizers on the growth and yield of wheat and barley. *Jour. Agric. South Australia* **35**, 57-86.
ROBERTS, S. M. (1964) *Dynamic Programming in Chemical Engineering and Process Control*. Academic Press, New York.

ROUMASSET, J. (1974) Estimating the risks of alternate techniques: Nitrogenous fertilization of rice in the Philippines. *Rev. Mktng. Agric. Econ.* **42**, 257–94.
ROUMASSET, J. (1976) *Risk and the Efficiency of Peasant Agriculture: The Case of Philippine Rice Farmers.* North-Holland, Amsterdam.
RUNGE, E. C. A. (1968) Effects of rainfall and temperature interactions during the growing season on corn yield. *Agronomy Jour.* **60**, 503–7.
RUSSELL, J. S. (1968a) Nitrogen fertilizer and wheat in a semi-arid environment: Part 2. Climatic factors affecting response. *Australian Jour. Expt. Agric. Animal Husb.* **8**, 223–31.
RUSSELL, J. S. (1968b) Nitrogen fertilizer in a semi-arid environment: Part 3. Soil and cultural factors affecting response. *Australian Jour. Expt. Agric. Animal Husb.* **8**, 340–8.
RUSSELL, J. S. (1968c) Nitrogen fertilizer and wheat in a semi-arid environment: Part 4. Empirical yield response models and economic factors. *Australian Jour. Expt. Agric. Animal Husb.* **8**, 736–48.
RUSSELL, J. S. (1972) A theoretical approach to plant nutrient response under conditions of variable maximum yield. *Soil Sci.* **114**, 387–94.
RYAN, J. G. (1972) A generalized crop-fertilizer production function. Ph.D. thesis, Dept. of Economics, North Carolina State Univ., Raleigh. (University Microfilms, Ann Arbor.)
RYAN, J. G. and PERRIN, R. K. (1973) The estimation and use of a generalized response function for potatoes in the Sierra of Peru. *North Carolina State Univ. Agric. Expt. Sta. Tech. Bul.* **214**, Raleigh.
RYAN, J. G. and PERRIN, R. K. (1974) Fertilizer response information and income gains: The case of potatoes in Peru. *Amer. Jour. Agric. Econ.* **56**, 337–43.
St. JOHN, R. C. and DRAPER, N. R. (1975) D-optimality for regression designs: A review. *Technometrics* **17**, 15–23.
SANDLAND, R. L. and JONES, R. J. (1975) The relation between animal gain and stocking rate in grazing trials: An examination of published theoretical models. *Jour. Agric. Sci., Camb.* **85**, 123–8.
SASIENI, M., YASPAN, A. and FRIEDMAN, L. (1959) *Operations Research.* Wiley, New York.
SATO, R. (1964) Diminishing returns and linear homogeneity: Comment. *Amer. Econ. Rev.* **54**, 744–5.
SCHLAIFER, R. (1959) *Probability and Statistics for Business Decisions.* McGraw-Hill, New York.
SEAGRAVES, J. A. (1970) Economic considerations in response research: Comment. *Amer. Jour. Agric. Econ.* **52**, 607–9.
SHARPE, P. R. and DENT, J. B. (1966) Some economic aspects of the use of growth stimulants in the nutrition of the bacon pig. *Univ. of Reading Farm Mgmt. Study No. 1*, Reading.
SHARPE, P. R. and DENT, J. B. (1968) The determination and economic analysis of relationships between plant population and yield of main crop potatoes. *Jour. Agric. Sci., Camb.* **70**, 123–9.
SHAW, L. H. (1964) The effect of weather on agricultural output. *Jour. Farm Econ.* **46**, 218–30.
SHRADER, W. D., FULLER, W. A. and CADY, F. B. (1966) Estimation of a common nitrogen response function for corn in different crop rotations. *Agronomy Jour.* **58**, 397–401.
SINDEN, J. A. (1964–5) An economic analysis to aid the marginal decision on rotation length. *Forestry* **37**, 161–78; **38**, 201–17.

REFERENCES   193

SINGH, I. and DAY, R. H. (1974) Estimating single nutrient yield response surfaces for new varieties with limited data. *Indian Jour. Agric. Econ.* **29**, 1–19.
SMITH, E. J. (1965) Technology in broiler production: impact on feed conversion and marketing weight. *U.S.D.A. Econ. Res. Service* ERS-246, Washington.
SMITH, W. G. and PARKS, W. L. (1967) A method for incorporating probability into fertilizer applications. *Jour. Farm Econ.* **49**, 1511–5.
SNEDECOR, G. W. and COCHRAN, W. G. (1967) *Statistical Methods*, 6th edn. Iowa State Univ. Press, Ames.
SOUTHEE, E. A. (1925) The law of diminishing returns. *Agric. Gazette of N.S.W.* **36**, 837–48.
SPILLMAN, W. J. (1933) Use of the exponential yield curve in fertilizer experiments. *U.S.D.A. Tech. Bul.* 348, Washington.
SPILLMAN, W. J. and LANG, E. (1924) *Law of Diminishing Returns*. World Book Co., New York.
STAUBER, M. S. and BURT, O. R. (1973) Implicit estimate of residual nitrogen under fertilized range conditions in the Northern Great Plains. *Agronomy Jour.* **65**, 897–901.
STAUBER, M. S., BURT, O. R. and LINSE, F. (1975) An economic evaluation of nitrogen fertilization when carry-over is significant. *Amer. Jour. Agric. Econ.* **57**, 463–71.
STEMBERGER, A. P. (1957) Economic implications of using alternative production functions for expressing corn-nitrogen production relations. *North Carolina State Univ. Agric. Expt. Sta. Tech. Bul.* 126, Raleigh.
STEWART, J. I. et al. (1974) Functions to predict optimal irrigation programs. *Jour. Irrign. Drainage Divn., Amer. Soc. Civil Eng.* **100**, 179–99.
STEWART, J. I. and HAGAN, R. M. (1973) Functions to predict effects of crop water deficits. *Jour. Irrign. Drainage Divn., Amer. Soc. Civil Eng.* **99**, 421–39.
SUNDQUIST, W. B. and ROBERTSON, L. S. (1959) An economic analysis of some controlled fertilizer input–output experiments in Michigan. *Michigan State Univ. Agric. Expt. Sta. Tech. Bul.* 269, East Lansing.
SWANSON, E. R. (1963) The static theory of the firm and three laws of plant growth. *Soil Sci.* **95**, 338–43.
SWANSON, E. R. and TYNER, F. H. (1965) Influence of moisture regime on optimum nitrogen and plant population for corn: A game theoretic analysis. *Agronomy Jour.* **57**, 361–4.
TANGRI, O. P. (1966) Omissions in the treatment of the law of variable proportions. *Amer. Econ. Rev.* **56**, 484–93.
THEIL, H. (1971) *Principles of Econometrics*. Wiley, New York.
THROSBY, C. D. (1961) Fitting production functions to experimental data. *Rev. Mktng. Agric. Econ.* **29**, 112–47.
THROSBY, C. D. (1964a) Some dynamic programming models for farm management research. *Jour. Agric. Econ.* **16**, 98–110.
THROSBY, C. D. (1964b) Theoretical aspects of a dynamic programming model for studying the allocation of land to pasture improvement. *Rev. Mktng. Agric. Econ.* **32**, 149–81.
TINTNER, G. and MILLHAM, C. B. (1970) *Mathematics and Statistics for Economists*. Holt, Rinehart and Winston, New York.
TOLLINI, H. and SEAGRAVES, J. A. (1970) Actual and optimal use of fertilizer: The case of nitrogen on corn in eastern North Carolina. *North Carolina State Univ. Dept. of Econ. Rept.* No. 14, Raleigh.

TOWNSLEY, R. (1968) Derivation of optimal livestock rations using quadratic programming. *Jour. Agric. Econ.* **19**, 347–54.
TOWNSLEY, R. (1969) Optimal rations for pigs fed separated milk and grain: A comment. *Jour. Agric. Econ.* **20**, 357–9.
TRAMEL, T. E. (1957) Alternative methods for using production functions for making recommendations. *Jour. Farm Econ.* **39**, 790–4.
TRANT, G. I. and WINDER, J. W. L. (1961) Time opportunity costs in broiler production. *Canadian Jour. Agric. Econ.* **9**, 123–9.
TWEETEN, L. G. and HEADY, E. O. (1962) Short-run corn supply and fertilizer demand functions based on production functions derived from experimental data: A static analysis. *Iowa State Univ. Agric. Expt. Sta. Res. Bul.* 507, Ames.
ULVELING, E. F. and FLETCHER, L. B. (1970) A Cobb-Douglas production function with variable returns to scale. *Amer. Jour. Agric. Econ.* **52**, 322–6.
UMLAND, A. W. and SMITH, W. N. (1959) Use of Lagrange multipliers with response surfaces. *Technometrics* **1**, 289–92.
UPTON, M. and DALTON, G. (1976) Linear production response. *Jour. Agric. Econ.* **27**, 253–6.
VAN MOESEKE, P. (1965) Diminishing returns and linear homogeneity: Comment. *Amer. Econ. Rev.* **55**, 536–9.
VALDÉS, A. (1967) Analisis economico de 20 ensayos de aplicacion de fertilizantes en trigo, maiz y papas. In E. Montero and S. Perez (Eds.), *Investigacion Economica y Experimentacion Agricola*, Instituto Interamericano de Ciencias Agricolas-Zona Sur-OEA, Montevideo, pp. 79–129.
VINOD, H. D. (1968) Econometrics of joint production. *Econometrica* **36**, 322–36.
VOSS, R. and PESEK, J. (1962a) Generalization of yield equations in two variables: Application of yield data from 30 initial fertility levels. *Agronomy Jour.* **54**, 267–71.
VOSS, R. and PESEK, J. (1962b) Estimation of effect coefficients relating soil test values and units of added fertilizer. *Agronomy Jour.* **54**, 339–41.
VOSS, R. and PESEK, J. (1967) Yields of corn grain as affected by fertilizer rates and environmental factors. *Agronomy Jour.* **60**, 567–72.
VOSS, R., HANWAY, J. J. and FULLER, W. A. (1970) Influence of soil, management and climatic factors on the yield response by corn (*Zea Mays* L.) to N, P, K fertilizer. *Agronomy Jour.* **62**, 736–40.
WALKER, O. L., HEADY, E. O. and PESEK, J. (1964) Application of game theoretic models to agricultural decision making. *Agronomy Jour.* **56**, 170–3.
WAUGH, D. L., CATE, R. B. and NELSON, L. A. (1973) Discontinuous models for rapid correlation, interpretation and utilization of soil analysis and fertilizer response data. *North Carolina State Univ. Soil Fertility Evaluation and Improvement Project Tech. Bul.* 7, Raleigh.
WHITMORE, G. A. (1970) Third-degree stochastic dominance. *Amer. Econ. Rev.* **60**, 457–9.
WILLIAMS, R. J. and BAKER, J. R. (1968) A comparison of response surface and factorial designs in agricultural research. *Rev. Mktng. Agric. Econ.* **36**, 165–77.
WILLS, I. R. and LLOYD, A. G. (1973) Economic theory and sheep-cattle combinations. *Australian Jour. Agric. Econ.* **17**, 58–67.
WINDER, J. W. L. and TRANT, G. I. (1961) Comments on determining the optimum replacement pattern. *Jour. Farm Econ.* **43**, 939–51.
WINDSOR, J. S. and CHOW, V. T. (1971) Model for farm irrigation in humid areas. *Jour. Irrign. Drainage Divn., Amer. Soc. Civil Eng.* **97**, 369–86.

REFERENCES

WINKLER, R. L. (1972) *Introduction to Bayesian Inference and Decision.* Holt, Rinehart and Winston, New York.

WRAGG, S. R. (1970) Co-operative research in agriculture and the provision of input/output coefficients. *Jour. Agric. Econ.* **21**, 85–98.

WRIGHT, A. and DENT, J. B. (1969) The application of simulation techniques to the study of grazing systems. *Australian Jour. Agric. Econ.* **13**, 144–53.

WU, I. and LIANG, T. (1972) Optimal irrigation quantity and frequency. *Jour. Irrign. Drainage Divn., Amer. Soc. Civil Eng.* **98**, 117–33.

YARON, D. (1971) Estimation and use of water production functions in crops. *Jour. Irrign. Drainage Divn., Amer. Soc. Civil Eng.* **97**, 291–304.

YARON, D. et al. (1963) Economic analysis of input–output relations in irrigation. In Intl. Comm. on Irrig. and Drainage, *Fifth Congress on Irrigation and Drainage*, New Delhi, pp. 16.13–34.

YARON, D. et al. (1972) Estimating procedures for response functions of crops to soil water content and salinity. *Water Resources Res.* **8**, 291–300.

YARON, D. et al. (1973) Wheat response to soil moisture and optimal irrigation policy under conditions of unstable rainfall. *Water Resources Res.* **9**, 1145–54.

YARON, D. and BRESLER, E. (1970) A model for the economic evaluation of water quality in irrigation. *Australian Jour. Agric. Econ.* **14**, 53–62.

YATES, F. (1949) The design of rotation experiments. *Cwlth. Bur. Soils Tech. Commun.* **46**, 142–55.

YATES, F. (1952) Principles governing the amount of experimentation in developmental work. *Nature* **170**, 138–40.

YATES, F. (1954) The analysis of experiments containing different crop rotations. *Biometrics* **10**, 324–46.

YATES, F. (1964) Sir Ronald Fisher and the design of experiments. *Biometrics* **20**, 307–21.

YATES, F. and BOYD, D. A. (1965) Two decades of surveys of fertilizer practice. *Outlook on Agriculture* **4**, 203–10.

YEH, M. H. (1964) The role of econometrics in agricultural research. *Canadian Jour. Agric. Econ.* **12** (2), 54–61.

ZELLNER, A. (1971) *Introduction to Bayesian Inference in Econometrics.* Wiley, New York.

ZELLNER, A. and REVANKAR, N. S. (1969) Generalized production functions. *Rev. Econ. Studies* **36**, 241–50.

Author Index

Abraham, T. P.   59, 60, 162, 169, 177
Agarwal, K. N.   162, 177
Alcantara, R.   60, 177
Allen, R. G. D.   12n, 14n, 26, 29n, 177
Anderson, J. R.   59, 60, 61, 99, 100, 143, 144, 145, 146, 150, 152, 157, 161, 164, 165, 168, 169, 170, 173, 174, 176, 177, 178, 180, 183
Anderson, R. L.   26, 61, 150, 158, 161, 169, 171, 172, 178
Anderson, T. W.   162, 178
Arcus, P. L.   99, 178
Arrow, K. J.   29n, 178
Arscott, G. H.   84n, 97, 180
Asce, M.   97, 185

Baird, B. L.   60, 178
Baker, J. R.   150, 194
Balaam, L. N.   150, 178
Balmukand, B. H.   26
Barker, R.   60, 144, 145, 168, 182, 190
Barrow, N. J.   98, 178
Bartholomew, W. V.   172, 178
Bath, J. G.   159, 186
Battese, G. E.   98, 99, 145, 159, 160, 162, 165, 178, 184, 185
Baule, B.   26
Baum, E. L.   149, 179
Beattie, B. R.   161, 180
Bennett, D.   99, 159, 179
Beringer, C.   98, 179
Bernsten, R. H.   174, 187
Bicking, C. A.   150, 179
Bie, S. W.   174, 179
Bilas, R. A.   12n, 179
Blackie, M. J.   168, 179
Blackman, F. F.   170, 179
Blandford, D.   146, 179

Blaxter, K. L.   26, 61, 179
Blight, B. J. N.   163, 179
Bofinger, V. J.   150, 162, 179
Bondavalli, B.   168, 179
Bondorff, K. A.   26
Boulding, K. E.   96, 179
Box, G. E. P.   26, 161, 163, 179
Box, M. J.   150, 180
Boyd, D. A.   26, 61, 150, 151, 153, 157, 163, 164, 172, 174, 180, 195
Bradley, R. A.   150, 161, 180
Bresler, E.   168, 195
Brody, S.   26, 180
Broeshart, H.   26, 184
Brown, W. G.   84n, 97, 145, 161, 167, 180, 190
Buras, N.   97, 98, 185
Burley, H. T.   93n, 99, 159, 183
Burt, O. R.   97, 98, 168, 180, 184, 193
Butcher, W. S.   97, 185
Byerlee, D. R.   145, 168, 180

Cady, F. B.   98, 159, 162, 163, 169, 180, 185, 189, 192
Campbell, N. A.   98, 175, 178, 180
Candler, W. V.   60, 84n, 174, 180, 186
Cannon, D. J.   99, 180
Carley, D. H.   60, 180
Carruthers, I. D.   144, 181
Carter, H. O.   26, 60, 98, 181, 184, 186
Carter, W. H.   59, 190
Castle, E. N.   84n, 99, 190
Cartwright, W.   60, 180
Cate, R. B.   61, 98, 170, 171, 172, 176, 181, 194
Chew, V.   26, 181
Chisholm, A. H.   97, 98, 99, 144, 145, 181

197

AUTHOR INDEX

Chow, V. T.   169, 194
Chowdhury, S. R.   163, 181
Clyde, H. S.   26, 181
Cochran, W. G.   150, 157, 160, 163, 181, 193
Colwell, J. D.   59, 60, 98, 145, 150, 161, 163, 167, 168, 169, 174, 181
Colyer, D.   145, 168, 179, 181
Cone, B. W.   144, 182
Conniffe, D.   99, 159, 182
Cowlishaw, S. J.   159, 182
Cox, G. M.   150, 157, 160, 163, 181
Currie, M.   146, 179

de Datta, S. K.   60, 174, 182, 187
de Janvry, A.   26, 60, 144, 145, 167, 182
de Oliveira, A. J.   60, 166, 182
Dalton, G.   26, 29n, 172, 194
Dano, S.   26, 59, 60, 182
Davidson, B. R.   174, 175, 182
Davis, J. F.   60, 182
Day, R. H.   145, 161, 182
Dean, G. W.   60, 84n, 98, 143, 144, 145, 160, 162, 163, 181, 182, 186
Decker, W. L.   168, 184
Denmead, O. T.   145, 182
Dent, J. B.   60, 61, 84n, 98, 100, 160, 169, 174, 182, 183, 192, 195
Dhrymes, P. J.   162, 183
Diamond, P. A.   144, 183
Dillon, J. L.   25, 26, 59, 60, 93n, 97, 99, 100, 143, 144, 145, 149, 150, 152, 157, 159, 161, 162, 164, 165, 168, 169, 170, 173, 176, 178, 181, 183, 186, 189
Doll, E. C.   145, 168
Doll, J. P.   26, 59, 60, 144, 145, 165, 168, 183, 184, 190
Donaldson, G. F.   144, 181
Dorfman, R.   43n, 60, 98, 183
Dowdle, B.   145, 183
Downey, L. A.   168, 183
Draper, N. R.   59, 150, 161, 180, 183, 192
Dudley, N. J.   97, 98, 168, 184
Duloy, J.   59, 99, 160, 162, 184
Dyckman, T. R.   144, 184

Eidman, V. R.   98, 184
Elandt, R. C.   150, 184
Engelstad, O. P.   60, 145, 168, 184
English, P. R.   60, 160, 183
Esdaile, R. J.   59, 168, 181
Evenson, R. E.   173, 184
Ewalt, R. L.   168, 184
Ezekiel, M.   161, 184

Faris, J. E.   96, 97, 98, 184
Fawcett, R. H.   61, 98, 172, 184
Federer, W. T.   150, 184
Finney, D. J.   173, 184
Fisher, R. A.   168, 184
Fitts, J. W.   168, 184
Fitzpatrick, E. A.   145, 168, 169, 190
Fletcher, L. B.   26, 194
Flinn, J. C.   97, 98, 184
Fox, K. A.   161, 184
Frakes, M. G.   60, 182
Freund, R. J.   97, 100, 162, 187
Fricke, E. F.   26, 191
Fried, M.   26, 184
Friedman, L.   97, 192
Frisch, R.   7n, 26, 59, 60, 96, 184
Fuller, W. A.   98, 145, 159, 161, 162, 165, 178, 185, 192, 194

Gaffney, M. M.   96, 97, 98, 185
Gallant, A. R.   161, 185
Gane, M.   96, 185
Gastal, E.   60, 185
Gillespie, R. H.   150, 179
Gnanadesikan, R.   162, 185
Godden, D. P.   26, 60, 61, 98, 185, 187
Graham, N. McC.   61, 179
Greig, I. D.   100, 185

Hadar, J.   97, 144, 185
Hadley, G.   165, 185
Hagan, R. M.   168, 193
Hall, W. A.   97, 98, 185
Hall, W. B.   162, 185
Halter, A. N.   26, 143, 144, 145, 163, 186
Hamilton, D.   159, 186
Hansen, P. L.   97, 186
Hanway, J. J.   145, 194

AUTHOR INDEX 199

Hardaker, J. B. 143, 144, 146, 165, 168, 173, 178
Harlan, J. R. 99, 186
Hart, R. H. 99, 186
Hartley, H. O. 168, 186
Havlicek, J. 60, 165, 168, 170, 173, 186
Hazell, P. B. R. 146, 186
Heady, E. O. 12n, 25, 26, 59, 60, 61, 84n, 93n, 97, 98, 99, 100, 144, 145, 149, 150, 157, 159, 161, 162, 163, 169, 170, 176, 181, 186, 187, 194
Helyar, K. R. 26, 60, 61, 98, 185, 187
Henderson, J. M. 12n, 59, 96, 187
Henson, T. L. 163, 188
Herdt, R. W. 174, 187
Hiebert, L. D. 145, 187
Hildreth, C. G. 169, 187
Hildreth, R. J. 168, 187
Hill, W. J. 150, 163, 179, 187, 188
Hirshleifer, J. 173, 187
Hjelm, L. J. 145, 187
Hochman, E. 97, 98, 144, 162, 187
Hocking, J. G. 26, 186
Hocknell, K. E. 99, 174, 188
Hoel, P. G. 162, 187
Hoepner, P. H. 97, 100, 162, 187
Hoerl, A. E. 161, 187
Hoffnar, B. R. 60, 150, 165, 168, 174, 187
Hogg, H. C. 168, 188
Hoglund, C. R. 149, 188
Holder, J. M. 160, 188
Hoover, L. M. 60, 188
Howell, D. T. 97, 98, 184
Huang, D. S. 161, 162, 188
Hunter, J. S. 150, 161, 180
Hunter, W. G. 163, 179, 187, 188
Hutton, R. F. 60, 188

Inkson, R. H. E. 167, 188
I.R.R.I. 145, 168, 188

Jackson, T. L. 167, 180
Jardine, R. 60, 62, 188
Jebe, E. H. 165, 183
Jensen, D. 166, 188
Jensen, E. 26

Johnson, G. L. 60, 150, 168, 174, 187
Johnson, N. L. 150, 157, 161, 188
Johnson, P. R. 169, 188
Johnson, R. W. M. 163, 188
Jones, R. B. 99, 174, 188
Jones, R. J. 26, 61, 99, 159, 188, 192
Jonsson, L. 26, 169, 188

Kaminsky, M. 162, 188
Keay, J. 176, 180
Kennard, R. W. 26, 161, 187, 188
Kennedy, J. O. S. 97, 98, 162, 188, 189
Kenworthy, O. O. 150, 189
Kislev, Y. 173, 184
Kmenta, J. 161, 162, 189
Knetsch, J. L. 145, 168, 189, 191
Kroth, E. M. 145, 168, 179, 182
Kupper, L. L. 150, 189

Laird, R. J. 163, 169, 180, 189
Lang, E. 26, 193
Lee, I. M. 97, 98, 144, 162, 187
Leftwich, R. H. 12n, 189
Leone, F. C. 150, 157, 161, 188
Liang, T. 168, 195
Liebig, J. von 26, 61, 170, 189
Lin, C. Y. 173, 189
Lindley, D. V. 165, 189
Lingle, J. C. 98, 184
Linse, F. 97, 98, 193
Lloyd, A. G. 99, 159, 189, 194
Lovell, A. C. 162, 189
Lucas, H. L. 159, 191
Lucas, J. M. 150, 189

McAdams, A. K. 144, 184
McArthur, I. D. 144, 145, 189
McCall, J. J. 143, 189
McCarthy, W. O. 144, 187
McConnen, R. J. 99, 189
McCorkle, C. O. 60, 181

Magnusson, G. 143, 189
Mandac, A. M. 145, 168, 189
Marquardt, D. W. 161, 189

Martin, B. R.   174, 175, 182
Mason, D. D.   26, 60, 178, 189
Mauldon, R. G.   99, 174, 182, 190
Mendenhall, W.   150, 160, 190
Merril, M. O.   145, 180
Meydrech, E. F.   150, 189
Mighell, R. L.   97, 186
Mihram, G. A.   169, 189
Milhorn, H. T.   61, 163, 189
Miller, R. L.   161
Millham, C. B.   193
Minhas, B. S.   98, 168, 190
Mitscherlich, E. A.   26
Montaño, C. B.   144, 145, 168, 190
Montero, E.   60, 190
Moore, C. V.   98, 190
Morley, F. H. W.   159, 190
Mott, G. O.   159, 191
Mundlak, Y.   162, 190
Munson, R. D.   26, 165, 183, 190
Musgrave, W. F.   97, 98, 184
Myers, R. H.   59, 150, 155, 157, 161, 190

Nagadevara, V.   163, 181
Natl. Res. Council (U.S.A.)   26, 61, 190
Needham, P.   61, 163, 172, 180
Neeley, D.   174, 187
Nelder, J. A.   26, 190
Nelson, L. A.   26, 61, 98, 150, 158, 161, 169, 170, 171, 172, 176, 178, 194
Nelson, M.   84n, 99, 190
Newman, P.   12n, 190
Nix, H. A.   145, 168, 169, 190
Norman, D. W.   60, 191

O.E.C.D.   61, 98, 150, 190, 191
Officer, R. R.   164, 183
Ogunfowora, O.   60, 190
Ostle, B.   160, 191
Ott, L.   163, 179
Oury, B.   145, 191
Owen, J. B.   159, 191

Paltridge, G. W.   100, 191
Parikh, K. S.   98, 168, 190
Paris, Q.   98, 191

Parks, W. L.   145, 168, 184, 191, 193
Pequignot, R.   145, 191
Perez, S.   60, 190
Perrin, R. K.   60, 61, 97, 145, 163, 165, 166, 167, 168, 170, 171, 172, 173, 174, 191
Pesek, J.   26, 60, 144, 145, 163, 166, 167, 168, 186, 187, 188, 191, 194
Petersen, R. G.   159, 167, 180, 191
Prato, A. A.   60, 177
Pratt, J.   173, 191
Prescott, J. A.   26, 191

Quandt, R. E.   12n, 59, 96, 187

Raeburn, J. R.   60, 183
Raghavarao, D.   150, 184
Raiffa, H.   165, 173, 191
Ram, M.   98, 191
Rao, P.   161
Rao, V. Y.   60, 167, 169, 177, 187, 191
Recamier, A.   145, 191
Reid, G. K. R.   100, 145, 162, 169, 191
Revankar, N. S.   26, 195
Richardson, A. E. V.   26, 191
Ridgman, W. J.   159, 191
Roberts, S. M.   97, 191
Robertson, L. S.   60, 193
Roumasset, J.   144, 145, 146, 168, 192
Runge, E. C. A.   168, 192
Russell, J. S.   26, 60, 145, 167, 192
Russell, W. R.   144, 185
Ryan, J. G.   26, 60, 145, 161, 163, 165, 166, 167, 168, 173, 192

St. John, R. C.   150, 192
Sandland, R. L.   26, 61, 99, 159, 188, 192
Sasieni, M.   97, 192
Sato, R.   29n, 192
Scandizzo, P. L.   146, 186
Schlaifer, R.   165, 173, 191, 192
Schneeberger, K. C.   169, 179
Seagraves, J. A.   60, 145, 163, 165, 168, 170, 173, 186, 192, 193
Sengupta, J. K.   167, 191
Sharpe, P. R.   60, 61, 192

Shaw, L. H.   145, 192
Shaw, R. H.   145, 182
Shrader, W. D.   98, 145, 159, 162, 178, 192
Sinden, J. A.   98, 192
Singh, I.   161, 193
Smidt, S.   144, 184
Smith, E. J.   97, 193
Smith, H.   161, 183
Smith, W. G.   145, 193
Smith, W. N.   59, 194
Snedecor, G. W.   150, 160, 193
Southee, E. A.   26, 193
Spedding, C. R. W.   159, 190
Spillman, W. J.   26, 193
Srinivasan, T. N.   98, 168, 190
Stackhouse, K. M.   161, 181
Stauber, M. S.   97, 98, 180, 193
Stemberger, A. P.   169, 193
Stewart, J. I.   168, 193
Stiglitz, J. E.   144, 183
Sundquist, W. B.   60, 182, 193
Swanson, E. R.   26, 61, 144, 170, 193

Tangri, O. P.   60, 193
Theil, H.   162, 193
Thomas, D. A.   100, 145, 162, 169, 191
Thorne, D. W.   60, 188
Throsby, C. D.   97, 99, 161, 193
Tintner, G.   26, 193
Tollini, H.   145, 163, 193
Tong Kwong Yuen, L.   61, 163, 172, 180
Townsley, R.   60, 98, 162, 194
Tramel, T. E.   59, 194
Trant, G. I.   96, 97, 194
Tweeten, L. G.   60, 186, 194
Tyner, F. H.   144, 193

Ulph, A.   174, 179
Ulveling, E. F.   26, 194
Umland, A. W.   59, 194
Upton, M.   26, 29n, 61, 172, 194

Van Moeseke, P.   29n, 194
Valdés, A.   60, 167, 194
Vinod, H. D.   162, 194
Voss, R.   145, 166, 167, 168, 194

Wainman, F. W.   61, 179
Walker, O. L.   144, 194
Waugh, D. L.   61, 98, 170, 171, 176, 194
Wheeler, J. L.   150, 179
Whitmore, G. A.   144, 194
Williams, R. J.   150, 160, 188, 194
Wills, I. R.   159, 194
Wilson, B. R.   160, 188
Winder, J. W. L.   96, 97, 194
Windsor, J. S.   169, 194
Winkler, R. L.   165, 194
Wragg, S. R.   61, 195
Wright, A.   100, 195
Wu, I.   168, 195

Yaron, D.   60, 98, 168, 195
Yaspan, A.   97, 192
Yates, F.   159, 160, 162, 164, 173, 174, 195
Yeh, M. H.   161, 195

Zellner, A.   26, 163, 195

Subject Index

Actuarial formulae 73
Ad libitum feeding 160
Adaptive research 150
Aggregation 163
Agricultural policy 146
Agricultural scientists 169
Amortization formula 74, 97
Analysis of variance 151, 160, 162
Animal decision making 83, 87, 90, 162
Animal liveweight 161
Annuity formula 74, 97
Attitude to risk 110, 111, 116
Autocorrelation 161
Available fertilizer 94
Average net revenue 73
Average product 5, 6, 23
Average profit per unit of time 71
Axioms of choice 107–8

Bayesian procedures 163
Beans 175
Beef 39
  feedlots 80, 98
Behavioural analysis 103
Berries 81
Best operating conditions 30–3, 34, 36, 38, 40, 41, 42–3, 55, 59, 60, 67, 131, 151, 172
  and differential calculus 59, 68
  and dynamic programming 68
  for grazing systems 86, 92, 99
  for sequential processes 76
  over time 70–7, 93, 97
  under constraints 48, 52, 56, 58
  under risk 102, 105, 107, 111, 113, 117, 121–2, 124, 131–3, 163, 176
Biological research 61
Biology of response 14, 26, 167, 169

Body weight 84, 99
Bone 40
Boundary solutions 45, 47, 49, 54, 58, 124
Broilers 65, 66, 79, 80, 83–6, 97–8, 144, 162
  objective function 85
  response model 84–5
  space requirements 85, 97–8
Butterfat 59

Calories 84
Canonical correlation 162
Cantaloupes 81
Capital 70
  intensity 174
Carbohydrate 14, 65, 83, 84
Carry-over effects 66, 98
  analysis 93–6
Cattle 80
Central composite design 152, 154–5, 158
Central treatment 154
Centre of design 158
Cereals 40, 79, 172
Certainty equivalent 129, 130
Choice axioms 107–8
Climatic variable 104, 167
  composite indices 168
  effects 126
  measurement 168
  sequences 168
  variability 145
Cobb–Douglas response function 24, 28, 40
Coding of factor levels 153–7
Commercial recommendations 174–6
Complete factorial design 152, 158

204 SUBJECT INDEX

Composite design 152, 154–5, 158
Compounded direct marginal cost 75
Compounding 67, 73
Computation 169
Computerized aids 175
Confidence limits 165
Conserved fodder 87
Conserved pasture 87, 90, 92
Constant prices 100
Constant returns 3
  to scale 11
Constrained maximization 44, 48, 54, 59, 77, 122–4
Constraints 32, 44
  effectiveness 54–5, 58
  functional 59
  least-cost 44
  on objective function 44, 48, 54, 57, 59, 77, 123
  on outlay 44, 59
  on output 44, 55
  on profit 77
  over time 77–8
  under risk 122–4
Continuity axiom 108, 129
Continuity of response 79
Continuous harvesting 66
Continuous planning 68
Continuous rate of interest 73, 125
Controlled input factors 42, 104–6, 135
Conversion factors 33
Co-operation in research 61
Correlation of yield and price 123, 126–7
Cost 32, 70, 85, 106, 113, 120
  functions 60
  locus 54, 56, 58, 77
  of credit 55
  per response period 71, 74
Co-third moments 134
Cows 83
Crop
  experiments 152, 159, 162
  fallow 96
  maturation pattern 81
  multi-harvest 79, 81–3, 98
  production 66, 79, 83, 93
  response 25, 26, 64, 78, 167
  response analysis 26
  response studies 60
  response to fertilizer 60, 80, 166, 170, 172
  response to irrigation 168
  response variability 165–9
  rotation 98, 159, 161–2
Cross-section data 161–2
Cubic utility function 133
Cucumbers 81
Cult of the asterisk 164
Cumulative input 65
Cumulative probability distribution 137, 139

Dairy production 83, 162
Decision theory 103, 143, 145, 173
Decision variables 104, 120, 163, 166, 167
  interaction 105
Decreasing returns to scale 3, 4, 7, 11, 43n
Degree of belief 109
Degree of preference 109
Demand for inputs 60, 66
Design of experiments 157
  for field trials 157–9
  for pen-feeding trials 159–60
Deterioration of pasture 90
Dietary pattern 65
Differential calculus 59, 68
Difficulties in response research 149
Digestible protein 83, 84
Diminishing returns 2, 4, 6, 7, 11, 24, 43n
Direct marginal cost 72, 75, 117
Discontinuous response 3
Discounting 67
Disease 145, 167, 168
Distribution of yield 105
Dynamic programming 68, 93–6, 97, 98, 99

Economic significance 163–5
Economics of experimentation 151–2, 158
Economics of response research 172–3
Efficiency 31, 64, 102
Egg production 83

SUBJECT INDEX 205

El Llano response trial 18, 22, 47, 50, 56
Elasticity
 of production 22
 of response 5, 7, 11, 26
 of substitution 11, 12, 22, 23
Empirical response studies 60
Environmental variables 145
Error mean square 165
Estimation of response function 151, 157, 160–5
Expansion path 46, 50
Expected price 115
Expected profit 103, 113, 144, 170
Expected utility 103, 107–9, 164
 Taylor series expansion 144
 theorem 107, 144
Expected utility maximization 103, 109–10, 113–14, 118, 131, 143, 173
 analytical approach 131, 167
 gross approach 131, 167
 moment method 131
 second-order conditions 118, 143
Expected yield 115, 166
 function 168
Experiment location 168
Experiment map 154
Experimental design 150, 152–7
 centre 158
 criteria 157–9
Experimental error 156
Experimental points 152–5
Experimental range 19
Experimental unit 157
Extension 149, 173

Factor substitution 11, 12n
Factor–factor relations 5, 11, 43, 50, 51, 58
Factor–product relations 5, 11, 43, 50, 51, 58
Factorial design 152–3, 158
Factors of production 1, 4, 157–8
 fixed 6, 10, 31n, 33, 64, 65, 67, 70
 levels 157, 158
 limiting 170–1
 range 158
 unimportant 4
 variable 4

Fallow 96
Farm
 extension 149, 173
 recommendations 175–6
 response 174–5
 survey data 174
 trials 174–5
Fat 40
Fat lamb production 83
Faustmann criterion 75, 96, 98
Feed 59, 164, 169
 ad libitum 99, 160
 constituents 83
 consumption 65, 84, 90
 cost 84
 demand 87
 input sequence 83
 mix 160
 sub-*ad libitum* 99
 supplement 87
 supply 65, 87
 types 83
 utilization 65, 84
Feeding period 83
Feedlot fattening 80, 98
Fertilizer 41, 81, 98, 164, 169, 172–3
 carry-over 79, 93, 97
 policy 94
 residual 94, 98
 response 18, 47, 61, 80, 98
 trials 152, 157
First-degree polynomial 27
Fishmeal 83, 84
Fixed assets 97
Fixed costs 70, 85, 106
Fixed input factors 4, 6, 10, 31n, 33, 64, 65, 67, 70, 85
 replacement 83
Fixed-outlay constraint 52, 56, 59
Fixed-output constraint 44, 48
Fixed prices 59
Fixed total return 44, 48
Floor space 85–6, 97–8
Flow of profit 68, 74, 125
Fodder conservation 87
Forestry 67, 75, 98
Form-free response function 169
Fractile rule 137

206　SUBJECT INDEX

Fractional factorial design　152, 153, 158
Future profit　73

Gain　17, 31, 67, 71
Game theory　103, 144
Generalized power function　28, 167
Generalized quadratic model　170–2
Generalized yield equation　166
Goodness of fit　169
Grain　40, 80, 160
Graphical procedures　77, 83, 119–20, 130, 137, 141
Grazing　86, 100
　best operating conditions　86, 92, 99
　constraints　87
　model　88–91
　objective function　91
　production　83, 86
　system　86, 88, 92–3
　trials　152, 159
Gross margin　$33n$, 111
　sensitivity　60, 62
Gross revenue　107
Growth pattern of crop　81

Harvest　64, 78, 80
　continuity　66
　intensity　79
　multiplicity　79, 81
　sequence　65, 66, 79, 86, 92
Hay　64, 79, 80, 87, 161
Hens　83
Historical cost　67
History of response analysis　25
Hyperbolic response function　169

Increasing returns　3, $43n$
　to scale　11
Independence axiom　108
Independent response processes　42, 48, 56, 57, 122–4
Industrial experimentation　150, 155
Initial soil nutrients　167
Input 1
　carry-over　66, 79, 93–6, 97, 98
　cumulation over time　65
　demand function　60, 66

　fixed　4, 6, 10, $31n$, 33, 64, 65, 67, 70
　injection pattern　79
　losses　17, 31, 67
　non-variable　5
　price uncertainty　106
　range　158
　replacement　96
　risk reducing　116–17, 168
　sequencing　65, 68, 86, 98
　supply function　65
　unimportant　4
　utilization equations　65
　variable　4
Input injections
　multiple　79
　single　79
Input–output function　162
Insecticide　116, 168
Insensitivity of profit　60
Instantaneous response　64
Intensity of harvest　79
Interest rate　67, 73, 125
Intermediate product　99
Inter-period climatic variability　145
Interrelated yield and price risk　126
Intra-period climatic variability　168
Invariable response period　79
Inverse price ratio　34, 37, 39
Investment criterion　172–3
Irrational input levels　34, 37
Irrigation　79, 97, 98, 116, 168
Isocline　11, 13, 24, 46, 58
　equation　13, 45, 50, 54
　shape　13
Iso-cost locus　54, 55, 56, 57, 58, 77
Iso-profit line　35
Isoquant　10, 13–16
　diagram　15, 20, 21, 48
　equation　11, 12, 45–6
　slope　14, $16n$
　surface　24, 25
Iso-revenue locus　49, 58
Isoutility curve　113–14, 115

Joint distribution
　of uncertain variables　167
　of yield and price　107, 134
Joint response　162

SUBJECT INDEX 207

Labour intensity 174
Lagrangian multiplier 44, 59, 62, 77, 123
Law of the minimum 61, 170
Least-cost expansion path 46, 49
Least-cost input array 44–6, 49
Least-cost ration 84n, 97
Least squares regression 160–1
Least-time ration 97
Level of significance 141, 152, 164–5, 176
Liebig's law 61, 170
Limited data 161
Limited outlay 52–8
Limiting factors 170
Linear programming 84n, 99
Linear response and plateau model 61, 161, 170–2
Livestock 162, 172
 bodyweight 84, 99
 decision making 83, 87, 90, 162
 experiments 152
 feed rations 83, 84n, 86, 97
 feeding trials 161
 maintenance 90–1
 minimal liveweight 88, 91
 non-grazing 83
 pen feeding 99
 production rate 90–1
 response 25, 26, 60, 64, 78, 84n, 97
 response studies 60
Livestock grazing 86, 100
 best operating conditions 86, 92, 99
 constraints 87
 model 88–91
 objective function 91
 production 83, 86, 99
 systems 93, 100
 trials 99, 152, 155, 159
Location of experiments 168
Logarithmic utility function 133
Loss 17, 31, 67
Lot-feeding 80, 99
Lump sums 67, 73

Maintenance restraint 91
Maize 83, 84
 fertilizer trial 157

Map of experiment 154
Marginal cost 69
 compounded 75
 direct 72, 75, 117
 of price variability 117
 of risk 115, 116, 118, 120, 128
 per unit of time 72, 75
Marginal expected product 115
Marginal expected revenue 117
Marginal net revenue 73
 over time 77
Marginal product 5, 6, 17, 23, 37
Marginal profit 60, 77
 per unit of time 71, 74
Marginal rate of substitution 12n
Marginal revenue 69, 73, 115
 per unit of time 75
Marginal risk 171
Marginal value product 69, 72
Marginal variance of revenue 117
Marginal variance of yield 115
Market price 33, 60
Mathematics of response 26
Mathematics of risk appraisal 143
Maximization of (expected) utility 103, 109–10, 113–14, 118, 131, 143, 173
 analytical approach 131, 167
 enumerative or gross approach 131, 167
 moment method 131
 second-order conditions 118, 143
Maximization of profit 32, 33, 35, 56, 58, 68
 over time 69–75, 96
Maximum average profit 71–2, 73
Maximum-cost input array 45, 47, 49, 52
Maximum financial profit 32–3
Maximum output 5, 7, 32, 171
Mean preserving spread 144
Mean price 115
Mean-skewness isoutility curves 128
Mean-variance frontier 114–15
Mean-variance profit space 113
Mean yield 115
Meat 40, 159, 162
Melons 81
Milk 59, 66, 79, 98, 160, 162

208 SUBJECT INDEX

Minimal liveweight 88, 91
Missing observations 161
Mitscherlich response function 28, 169
Model evaluation 170
Moment estimation formulae 136, 138
Moments of yield 136
Multicollinearity 161
Multi-equation model 85, 162
Multi-harvest crops 79, 81, 98
Multi-stage production 79, 86, 99
Multiple decision variables under risk 120–2
Multiple response 40, 42, 59, 162
Muscle 40

n variable inputs 23, 39
n variable response function 23
Net gain 31, 72
Net return 55
Net revenue
  average 73
  marginal 73, 77
Nitrogen response 172
Non-homogeneous error variance 163
Non-linear regression 161
Normative analysis 30, 60, 103
Number of experimental factors 151

Objective function 31, 33, 36, 39, 42, 59, 60, 67, 74
  constrained 44, 48, 54, 57, 77, 123
  for broilers 85
  for livestock grazing 91
  for multi-harvest crop 82
  in utility terms 109–11, 112, 113, 121, 125
  over time 66, 68, 70
  under risk 103, 106, 109, 121, 125, 127
  with input carry-over 94
One-sixth fractional factorial 153
Opportunity cost 55, 59, 66, 75, 94, 173, 176
  of time 69, 72, 97
Optimal grazing system 93
Optimal policy 94–6
Orcharding 98

Ordering axiom 107
Orthogonal polynomial 161
  trend 168
Outlay constraint 51, 55, 57–8, 59, 77
Output 1
  constraint 44, 55
  contours 10
  gains 31, 67
  harvests 65, 79
  maximum 5, 32
  supply functions 60

Pasture 76, 79, 83, 87, 90, 99
  conservation 87, 89, 91, 92
  deterioration 90–1
  grazing 86–93, 159
  hay 64, 79
  production 90
  stocking rate 99, 144, 152, 159
Pasture-livestock complex 89–92, 99
Pen-feeding trials 99, 159–60
Perennial crops 161
Pests 145, 167, 168
Physical mechanism of response 26, 162
Pigs 98, 160, 162, 172
Pirque response trial 50
Planning 68
Plot size 151
Polynomial response function 155–6, 167
  first-degree 27
  generalized 170–2
  orthogonal 161
  second-degree 28, 40, 156
  spliced 161
  square-root 169–70
Pork costulator 176
Positive analysis 30
Potatoes 41, 172, 173
Power response function 24, 26, 28, 40, 169
Power utility function 110, 133
Predetermined variables 104–5, 135, 165, 167, 171
Preferences 100, 102, 107–9, 111, 139
Present value 67, 73–4, 97
Price 33, 60
  certainty 100

SUBJECT INDEX 209

fixed 59
non-constant 59
policy 106
ratio 34, 37, 38, 43, 50
risk 120, 122, 134, 137, 176
uncertainty 106–7
weights 60
Price-yield risk interaction 117, 120
Principle of limiting factors 170
Prior information 158, 163, 165
Probability
 judgements 176
 of loss 144
 specification 134–7
 subjective 103, 105
Probability distribution 103, 131
 cumulative 139–40
 of price 107
 of profit 107, 110, 134, 163
 of utility 134
 of yield 107, 135, 163
Product 1
 price distribution 107
 price uncertainty 106
Production function 1, 60
Product-product relations 5, 43, 49, 51, 58
Profit 33, 73
 distribution 107, 110, 163
 flow 68, 74, 125
 lump sum 125
 mean 111, 145, 173
 moments 111, 132
 per unit of time 70
 relevant range 130
 sensitivity 60
 skewness 111, 127–8
 stream 73
 uncertainty 102, 106–7
 variance 111, 145, 173
Profit maximization 32, 33, 35, 37, 56, 58, 59, 68, 71
 constrained 77
 over time 59–75, 96
 with input carry-over 94
Protein 14, 65
Purchased feed 92
 restraint 91
 sequence 88

Quadratic response function 28, 40, 167, 169–72
Quadratic utility function 116, 131, 132–3

Rainfall 104, 168
Range of outcomes 176
Rate
 of feed consumption 90
 of interest 67, 73, 125
 of livestock production 90
 of pasture deterioration 90
 of pasture production 90
 of stocking 87, 99, 144, 159
 of substitution in response 114
 of substitution in utility 113, 127
 of technical substitution 11–17, 23, 37
 of technical transformation 43
Ration lines 160
Rations for livestock 83, 84n, 86, 97
Recommendations to farmers 174, 175–6
Recurrence equations 94
Regional recommendations 176
Regression analysis 161
 coefficients 165
 least-squares 160–1
 non-linear 161
 ridge 161
Relative yields 170
Relevant range 37, 43n, 116
 of profit 130
Repeated observations 161
Replacement problem 83, 96–7
Replication 151, 158
Residual fertilizer 94, 98, 167
Resistance function 28
Response 1
 assumptions 3, 4
 biological aspects 167, 169
 coefficients 2
 concurrent 79–80
 constrained 59
 continuous 79
 crop-fertilizer 60, 80, 166, 170, 172
 curve 4
 efficiency 30, 64, 102
 elasticity 5, 7, 11, 26

SUBJECT INDEX

in experiments 174–5
instantaneous 64
linear 61, 161, 170–2
mean 118, 136
models 80, 169–72
multi-stage 86, 99
multiple 162
on farms 174–5
over time 174–5
parameters 2
period 71, 79, 83
physical mechanism 26
research difficulties 149
sequential 71, 81, 97
simulation 168
simultaneous 26, 41
single variable 4, 7
skewness 136
spatial effects 163, 165, 168
stages 79
surface 8, 10
time-dependent 70, 97
to nitrogen 172
variability 165–9
variance 118, 136
Response analysis
  difficulties 149
  empirical studies 26, 60
  history 25
  mathematical basis 26, 59, 169
  over time 80
  principles 150, 175
  purposes 30
Response function 1, 4, 10
  algebraic form 26, 27–9, 161, 162, 166, 169–71
  Cobb–Douglas 24, 28, 40, 167, 169
  curvature 60
  estimation 151, 157, 160–5
  expected 168
  first-degree polynomial 27
  form free 169
  generalized 166
  generalized polynomial 170–2
  generalized power 28, 167
  hyperbolic 169
  Mitscherlich 28, 169
  n variable 23
  orthogonal polynomial 161

polynomial 28, 155–6, 167
power 24, 26, 28, 40, 167, 169
quadratic 28, 40, 167, 169–72
resistance 28
second-degree polynomial 28, 40, 156
Spillman 28
spliced polynomial 161
square-root 169, 170
time-dependent 64
transcendental 28
two-part 171
Response model
  alternatives 151, 170
  comparisons 172
  evaluation 169–72
  for broilers 83–6
  for grazing livestock 86–92
  for multi-harvest crop 81–2
  generalized quadratic 170–2
  linear 170–2
  multi-factor 171–2
  with fertilizer carry-over 93–6
Response research 163
  difficulties 149
  economics 172–3
  investment criterion 172–3
Response surface 8, 10, 20, 159
  estimation 157, 160–1
  designs 152–7
Response theory 2–3, 7
  biological 26
  logic 26
  mathematics 26
Returns to scale 4, 11, 43n
Revenue per response period 71, 74
Rice 144, 145, 168, 175
Ridge-line 11, 13, 15–17, 24
Ridge regression 161
Risk 30, 100, 102, 113–19, 120–2, 144, 165, 171
  analysis 167
  and best operating conditions 112, 122
  attitudes 110, 111, 176
  aversion 110, 116, 140, 171
  effects 125
  efficiency analysis 140, 142, 144
  empirical analysis 129
  indifference 110, 116

preference   110, 116
underestimation   167
Risk-discount factor   103
Risk-reducing factors   116–17, 168
Risky choice   102
Risky response appraisal   144
Risky response problem   134
Risky time-sequences   144
Rotatable design   152, 155–7, 158, 160
Rotation experiments   159, 161–2
Rules of stochastic dominance   139–41

Safety-first rule   103, 144
Sample size   165
San Pablo response trial   41
Scale of farming   174
Scale of production   59
Scientific co-operation   61
Scientific error   164
Scientific objectivity   164
Seasonal variation   99
Second-degree polynomial   28, 40, 156
Second-order conditions   7, 26, 36, 39, 45, 49, 118, 143
Seed   83
Sequence
  of harvests   68, 79
  of inputs   68, 80
  of observations   161
  of response processes   70, 73–5, 79–80
Sequential best operating conditions   76
Series of experiments   163–5
Set-up period   82
Sheep   83, 162
Significance
  economic   163–5
  levels   151, 152
  statistical   163–5, 169, 176
  tests   164, 176
Significant differences   151
Simulation   93, 98, 100, 168, 172, 173
Simultaneous equations   162
Simultaneous response processes   26, 41
Single harvest   79
Single input injection   79
Single-stage process   79
Single variable input   4, 6, 7, 33
Sinking fund formula   74

Skewness   134
  effects   128
  of profit   127–8
  of yield   136, 145, 166
Soil
  characteristics   167
  classification   173–4
  moisture   104
  nutrient status   98, 165
  testing   167, 171, 172, 173–4
Solar energy   144, 145, 168
Sparse data   136, 144
  procedures   137
Spatial variability   163, 165, 168
Specification of probabilities   134–7
Spillman function   28
Spliced polynomial   161
Square-root response function   169, 170
Statistical dependence   126
Statistical estimation   160–1
Statistical significance   163–5, 169
Stochastic dominance   129, 138, 144
  first-degree   140
  second-degree   140
  third-degree   140
Stocking rate   144, 152, 159
  trials   99
Stomach capacity   99
Sub-*ad libitum* feeding   99
Subjective judgement   100, 102, 164, 169, 173
Subjective probability   103, 105, 107–9, 110, 134, 143, 145
Substitution between factors   14
  elasticity   11, 12
  marginal rate   12n
  rate   11, 12, 13–18
Sufficient statistics   165
Sugar beet   172
Sugar-cane   175
Supply of input   65–6
Supply of output   60
Survey data   174
Systems modelling   100
Systems of grazing   86

Taxation effects   60, 97
Taylor series   132, 144, 170

212 SUBJECT INDEX

Technical units 60
Temperature 104, 168
Tests of statistical significance 176
Theory 3
 of continuous response 2, 7
 of discontinuous response 3
 of linear response 170–2
Three-factor rotatable design 156
Threshold yield 171
Time 3, 30, 64, 70, 77, 102, 124, 163, 165, 168
 and best operating conditions 70–7, 93, 97
 and risk 124–6
 as variable input 64
 classification of response processes 78–80
 effects 64, 66
 of harvest 79
 of irrigation 168
 of replacement 83
 opportunity cost 69, 72, 75, 97
 preference 67, 73, 76, 94, 96
 series 135, 161–2
Tobacco 175
Tomatoes 81
Trace elements 3, 167
Traditional statistical procedures 164, 176
Transcendental response function 28
Treatments 152, 158
 central 154
 coding 152–4
 number 151
Trials
 crop-fertilizer 152, 159, 162, 174
 farm 174–5
 grazing 152, 159
 pen-feeding 99, 159–60
True response function 170
Type I error 164
Type II error 164

Uncertain variables in response 165
Uncertainty 67, 102, 166
 in choice 102
 in price 106
 in yield 106

Uncontrolled response variables 104, 134, 166
Unimportant input factors 4
Utility 107–12, 143
 analysis 171
 differential 113–14
 interpersonal comparison 109
 maximization 103, 109–10, 113–14, 118, 131, 143, 173
 objective function 109–10, 113, 121, 123
 scale 109
 theorem 107, 144
Utility function 108–12
 algebraic estimate 131
 algebraic form 110
 cubic 133
 graph 130
 logarithmic 110, 133
 moment form 113, 127
 positive linear transformation 109, 131
 power 110, 133
 quadratic 110, 111, 116, 131, 132–3
 specification 129–31
 Taylor series expansion 111, 132, 144
 uniqueness 109

Value of information 170, 172, 173
Variability of response 165–9
Variable costs 106
Variable inputs 4, 76
Variables
 climatic 104, 167
 controlled 104
 critical level 171
 decision 104–5, 120, 166–7
 environmental 145
 predetermined 104, 165–7
 uncertain 104, 165–6
 uncontrolled 104, 105, 134, 166
Variance
 non-homogeneous 163
 of price 115
 of profit 113
 of yield 115, 145, 166
Variance-covariance matrix 165
von Neumann-Morgenstern method 130
Vertical integration 86

Water stress 145
Weather predictions 145
Weight gain 159
Wheat 18, 47, 50, 56, 175
Wool 79, 80, 83, 162

Yield 1
  distribution 107, 135, 163
  farm vs. experimental 174–5
  insurance 105
  maximum 5, 7, 32, 171
  moments 136
  plateau 171
  risk 120, 121, 134, 176
  skewness 136, 145, 166
  threshold 171
  uncertainty 105
  variability 165–9
  variance 115, 145, 166

Zero treatment 154

5-9
8.50
2